Bindon Blood Stoney

The strength and proportions of riveted joints

Bindon Blood Stoney

The strength and proportions of riveted joints

ISBN/EAN: 9783337156602

Printed in Europe, USA, Canada, Australia, Japan

Cover: Foto ©Andreas Hilbeck / pixelio.de

More available books at **www.hansebooks.com**

THE
STRENGTH AND PROPORTIONS
OF
RIVETED JOINTS.

BY

BINDON B. STONEY, LL.D., F.R.S., M.R.I.A., M.R.D.S.,

MEMBER OF THE INSTITUTION OF CIVIL ENGINEERS AND OF THE INSTITUTION OF NAVAL ARCHITECTS;

ENGINEER TO THE DUBLIN PORT AND DOCKS BOARD.

E. & F. N. SPON, 125, STRAND, LONDON.
NEW YORK: 35, MURRAY STREET.
1885.

PREFACE.

THE subject of riveting is by no means so simple as might at first sight be supposed, and the author, having had much trouble in collecting and arranging the various experiments which have been published on the subject and drawing thence practical conclusions for his own guidance, thinks that other Engineers, who have neither time nor opportunity for traversing the same ground, may, perhaps, find the following paper a useful *résumé* of our present knowledge on the subject of riveting. It was originally read at the Institution of Civil Engineers in Ireland, and the author is indebted to the Council of the Institution for their kind permission to publish the paper in a separate form from the "Transactions."

CONTENTS.

PART I.

IRON PLATES AND IRON RIVETS.

ART.		PAGE
1.	Shearing strength of bar iron,	3
2.	Shearing strength of iron rivets in the joint,	5
3.	Size and pitch of iron rivets—Proportions of joints—Boilermakers' practice—Shipbuilders' practice,	11
4.	Tensile strength of perforated plates,	21
5.	Lap of plates and pitch of rivets,	24
6.	Lap joints, effect of bending,	26
7.	Single-riveted lap joints—Reduction of unit-strength of plate—Efficiency of joints,	28
8.	Double riveted lap joints,	31
9.	Butt joints—Crushing pressure of rivets,	32
10.	Contraction of rivets and resulting friction of plates,	34
11.	Bearing area of iron rivets,	38
12.	Strength of iron rivets in tension,	40
13.	Efficiency of riveted iron joints,	40
14.	Theoretic proportions of joints,	41

PART II.

STEEL PLATES AND STEEL RIVETS.

15.	Shearing strength of bar steel by direct experiments,	45
16.	Shearing strength of steel rivets in the joint,	46
17.	Tensile strength of perforated steel plates (not riveted),	55
18.	Tensile strength of plates in the joint,	63
19.	Margin and lap of plates,	73
20.	Friction of steel joints and slip of plates,	74
21.	Bearing pressure of rivets,	77
22.	Efficiency of steel joints,	78
23.	Proportions of joints—Boilermakers' practice—Shipbuilders' practice—Girderwork,	79
24.	Theoretic proportions of steel joints,	85

THE STRENGTH AND PROPORTIONS OF RIVETED JOINTS.

PART I.

IRON PLATES AND IRON RIVETS.

1. *Shearing Strength of Bar Iron.*—On the subject of shearing, Professor Unwin makes the following remarks:[*]— "In Wöhler's researches (in 1870) the shearing strength of iron was found to be $\frac{4}{5}$ of the tenacity. Later researches of Bauschinger confirm this result generally, but they show that for iron the ratio of the shearing resistance and tenacity depends on the direction of the stress relatively to the direction of rolling. The above ratio is valid only if the shear is in a plane perpendicular to the direction of rolling, and if the tension is applied parallel to the direction of rolling. The shearing resistance in a plane parallel to the direction of rolling is different from that in a plane perpendicular to that direction, and again differs according as the plane of shear is perpendicular or parallel to the breadth of the bar. In the former case the resistance is 18 to 20 per cent. greater than in a plane perpendicular to the fibres, or is

[*] Proc. I. M. E., 1881, p. 327.

equal to the tenacity. In the latter case it is only half as great as in a plane perpendicular to the fibres."

Table I. gives the shearing strength of iron derived from direct experiments on rolled bars. Clark's experiments were made with a bent lever, the fulcrum of which was a round bar in place of a knife edge, and it is probable that the apparent shearing strength in his experiments was in excess of the real strength, in consequence of the friction of the apparatus and the uncertainty as to the exact leverage.

TABLE I.—*Tensile and Shearing Strengths of Bar Iron.*

Authority.	Diameter of bar.	Tensile strength per sq. inch.	Shearing strength per square inch.		Observations.
			Single shear.	Double shear.	
	Inch.	Tons.	Tons.	Tons.	
Clark,	⅝	24	24·15	22·1	Rivet iron of excellent quality.
Greig & Eyth,	⅝	22·23	—	19·01	Taylor's Yorkshire rivet iron.
Shock,	½	—	19·68	18·32	Ordinary round bar of commerce.
,,	⅝	—	17·41	17·23	Do.
,,	¾	—	17·61	17·76	Do.
,,	⅞	—	18·50	16·88	Do.
,,	1	—	17·90	16·78	Do.
Harkort	—	26·4	16·5		Not stated whether in single or double shear.
Lavalley	—	25·4	20·2		Do.
Dick,	1	26·0	21		Rivet bars, Do.
,,	1	23·8	20		Do. Do.
,,	1	24·2	20·2		Do. Do.
,,	1	24·1	19·4		Do. Do.

(Shock: Mean single shear 18·22, Mean double shear 17·40. Dick: Mean single 24·5, Mean double 20·15.)

Clark; Britannia and Conway tubular bridges, p. 390.
Greig & Eyth; Proc. I. M. E., 1879, pp. 292, 293.
Shock; Clark's Manual of Rules, &c., for Mechanical Engineers, p. 587.
Harkort & Lavalley; Proc. I. M. E, 1881, p. 327.
Dick; Trans. Inst. of Eng. and Shipbuilders in Scotland, Vol. XXV., p. 67.

2. *Shearing Strength of Iron Rivets in the Joint.*—Professor Unwin has made a valuable report on riveted joints to the Research Committee of the Institution of Mechanical Engineers, from which, as well as from the original accounts of the experiments, Tables II., V., and VI. have been derived.*

TABLE II.—*Shearing Strength of Iron Rivets in Single Shear, derived from Experiments on* SINGLE-RIVETED LAP JOINTS, *with* DRILLED HOLES, *broken by the Rivets Shearing.*

Mode of riveting.	Tensile strength of solid plate, Tons per sq. in.	Stress in joint at moment of fracture, Tons per square inch.			Size and number of rivets.	Remarks and source of experiments.
		Tensile, on net plate area.	Shearing.	Crushing.		
Hand,	22·00	15·10	18·63 ⎫	29·63	five ⅞″	Stoney, ⅞″ plates.
,,	,,	17·75	17·90 ⎬ Mean, 18·28	28·84	Do.	Do. do.
,,	,,	20·90	18·30 ⎭	28·75	Do.	Do. do.
Machine,	—	—	18·51 ⎫	—	·84″	Fairbairn.
Hand,	—	—	20·34 ⎬ Mean, 19·48	—	·82″	Do.
Machine,	—	—	19·58 ⎭	—	·84″	Do. Countersunk.
Steam,	22·25	19·48	18·44 ⎫	26·56	—	Greig and Eyth.
,,	,,	19·63	18·61 ⎬ Mean, 18·43	26·77	—	Do.
Hydraulic,	,,	20·43	19·35 ⎪	27·86	—	Do.
Steam,	,,	21·29	17·31 ⎭	29·59	—	Do.
Mean shearing strength of rivets in single riveted lap joints with *drilled* holes,		18·70	—	—		—

* Proc. Inst. M. E., 1881, pp. 341 to 344. Trans. Roy. Irish Ac., Vol. XXV., p. 451. Proc. Roy. Soc., 1873, p. 261.

TABLE III.—*Shearing Strength of Iron Rivets in Single Shear, derived from Experiments on* SINGLE-RIVETED LAP JOINTS, *with* PUNCHED *Holes, broken by the Rivets Shearing.*

Mode of riveting.	Tensile strength of solid plate, Tons per sq. in.	Stress in joint at moment of fracture, Tons per square inch.			Size and number of rivets.	Remarks and source of experiments.
		Tensile, on net plate area.	Shearing.	Crushing.		
Hand,	22·00	11·97	17·78 (Mean, 18·84)	27·94	five $\frac{3}{4}''$ rivets.	Stoney, $\frac{3}{8}''$ plates.
,,	22·00	14·75	19·90	32·60	Do.	Do. do.
Machine,	—	—	19·53 (Mean, 20·45)	—	·84″	Fairbairn.
Hand,	—	—	20·61	—	·82″	Do.
,,	—	—	21·20	—	·82″	Do. Countersunk.
Mean shearing strength of rivets in a single riveted lap joint with *punched* holes,	19·80	—	—	—		

TABLE IV.—*Shearing Strength of Iron Rivets in Single Shear, derived from Experiments on* DOUBLE-RIVETED LAP JOINTS, *with* PUNCHED *Holes, broken by the Rivets Shearing.*[*]

Mode of Riveting.	Tensile strength of solid plate, Tons per sq. in.	Stress in joint at moment of fracture, Tons per square inch.			Size and number of rivets.	Remarks and source of experiment.
		Tensile, on net plate area.	Shearing.	Crushing.		
?	21·6	19·8	18·6	?	Four $1\frac{1}{8}''$ rivets.	Knight, $\frac{7}{16}''$ B. B. boiler plates.
Hand,	18·25	14·23	19·35	29·9	Two $\frac{3}{4}''$ rivets.	Stoney, $\frac{3}{8}''$ plates.
Mean shearing strength of rivets in double-riveted lap joints with *punched* holes,	18·98	—	—	—		

[*] Proc. Inst. M. E., 1881, p. 720. Trans. Roy. Irish Ac., Vol. XXV., p. 451.

TABLE V.—*Shearing Strength of Iron Rivets in Single Shear, derived from Experiments on* SINGLE-RIVETED *and* SINGLE-COVERED BUTT Joints, *with* PUNCHED *Holes, broken by the Rivets Shearing.*

Mode of riveting.	Tensile strength of solid plate, Tons per sq. in.	Stress in joint at moment of fracture, Tons per square inch.			Size and number of rivets.	Remarks and source of experiment.
		Tensile, on net plate area.	Shearing.	Crushing.		
Steam, -	24·08	13·87	17·92	20·06	—	Kirkaldy, single cover.

Steel is harder than iron, and iron rivets are shorn by steel plates with a lower unit-stress than occurs with iron plates, in the ratio of $16\frac{1}{2}$ tons to $19\frac{1}{4}$ tons per square inch of rivet section, *i.e.*, about 14 per cent. less, according to Admiralty experience.*

Table VI. gives the results of several other experiments on the shearing strength of iron rivets connecting steel plates.†

Mr. Denny inferred from his experiments (Table VI.) that the shearing strength of iron rivets is 19 tons per square inch.‡ It will be observed that his experiments were made on *single* rivets in double shear, and the shearing unit-strength of a single rivet is probably somewhat greater than that of several rivets shorn together. In the discussion on Mr. Denny's paper, Mr. Parker says:—"In an iron joint the strength of the rivets to resist shearing will not probably be more than 18 or 19 tons per square inch,"§ and it will be observed in the experiments recorded by Mr. Parker (Table VI.) that the shearing strength per square inch of the small-sized rivets in $\frac{1}{4}$ inch plates was much greater than that of the larger rivets in the thicker plates.

* Trans. Inst. Nav. Arch., 1884, p. 274, and 1885, p. 189.
† Proc. Inst. M. E., 1881, p. 350. Trans. Inst. Nav. Arch., 1878, pp. 13, 14, and 1880, p. 204.
‡ Trans. Inst. Nav. Arch., 1880, p. 192.
§ *Idem*, p. 222.

TABLE VI.—*Shearing Strength of Iron Rivets in Joints of* STEEL *Plates, broken by Shearing.*

Mode of Riveting.	Holes made by.	Tensile strength of steel plate. Tons per sq. inch.	Stress in joint at moment of fracture, Tons per square inch.			Remarks and source of experiments.
			Tensile, on net plate area.	Shearing.	Crushing.	
Double-riveted lap, single shear,	Punch,	27·4	24·71	19·36	24·71	Knight.
,, ,,	,,	25·8	24·86	19·54	24·86	,, Plates annealed.
Treble-riveted lap, single shear,	Drill,	28·8	31·92	17·4	23·93	Parker, $\frac{7}{16}$" plate.
,, ,,	,,	27·6	26·42	16·7	20·80	,, $\frac{3}{4}$" plate.
,, ,,	,,	28·0	29·22	15·2	20·22	,, $\frac{3}{4}$" plate.
,, ,,	,,	26·7	28·16	15·9	20·12	,, $1\frac{5}{8}$" plate.
,, ,,	,,	30·0	25·55	16·5	20·10	,, $\frac{3}{4}$" plate.
,, ,,	,,	30·7	34·76	19·1	27·03	,, $\frac{1}{4}$" plate.
,, ,,	,,	32·2	34·95	19·2	27·12	,, $\frac{1}{4}$" plate.
Hydraulic, single rivet in double shear,	,,	—	—	19·3	—	Denny, mean of 4 experiments, $\frac{3}{4}$" iron rivets made from best scrap iron.
Hand, single rivet in double shear,	,,	—	—	18·7	—	Do. do.
Double-riveted butt, one cover, single shear, not annealed,	Punch,	—	—	16·7	—	Martell, $\frac{3}{4}$" rivets, and $\frac{1}{2}$" plate and cover.
,, ,, annealed,	,,	—	—	17·1	—	Do. do. do.
Double-riveted lap, single shear, not annealed,	,,	—	—	19·2	—	Do. $\frac{13}{16}$" rivets and $\frac{13}{32}$" plates.

Mean = 17·1.

Brunel made experiments on double-riveted and double-covered butt joints with punched holes, in one of which the rivets (in double shear) were sheared with a stress of 20·6 tons per square inch,* and, many years since, Mr. Doyne made experiments on riveted joints, no doubt with punched holes, and he states that the average shearing strength of rivets is 18·82 tons per square inch in single shear in a single-riveted lap joint, and 17·55 tons per square inch in double shear.†

Sir Edward Reed states, as the results of carefully conducted experiments at Chatham Dockyard, that the mean shearing strength of a $\frac{3}{4}$ inch rivet of Lowmoor or Bowling rivet iron in single shear was 10 tons, and when shorn in two places 18 tons. As, however, he does not give the exact diameters of the rivets after they filled the holes, these experiments do not throw much light on the subject, except to show that the shearing unit-strength of iron rivets is greater in single than in double shear.‡

From these various experiments, and having regard to the fact that riveting for experiments is probably done more carefully than the average run of actual work, we are not warranted in assuming higher standards for the shearing strength of iron rivets in single shear than 19 *tons per square inch in punched iron plates,* and **18**, or at most **18·5** *tons in drilled plates,* the rivets in the latter being weaker, probably, as Maynard suggests, because the sharp edges of drilled holes have a tendency to shear off the rivet cleaner than those of punched holes. Adopting 4 as the usual factor of safety, this makes the working shearing stress of iron rivets in iron plates from 4·5 to 4·75 tons per square inch, though 5 tons is the standard adopted by some engineers. Mr. Milton,

---

* Clark's Manual for Mech. Engineers, p. 640.
† Proc. Inst. C. E., Vol. IX., p. 357.
‡ Reed on Shipbuilding, p. 351.

one of **Lloyd's surveyors,** states that Lloyd's rules for boilers practically credit " rivets either in punched or drilled holes with a strength of 18 tons per square inch."* The calculations in the Board of Trade rules for marine boilers are apparently based on a shearing strength of 21 tons per square inch, but their minimum factor of safety is 5, and this does not allow the working shearing stress to exceed **4·25 tons per square inch.** Mr. Shaler Smith, the distinguished American engineer, adopts 4·46 tons per square inch for the working shearing stress of pins, bolts, and rivets.† Other American engineers, however, specify that the shearing stress of rivets and bolts shall not exceed 2·68 tons per square inch.‡ The balance of evidence seems to show that the shearing unit-strength of iron rivets is greater in single than in double shear, and probably also it is greater in double-riveted than in single-riveted lap joints. Lloyd's and the Board of Trade rules for marine boilers recognise this, and they direct that the shearing area of rivets in double shear shall be calculated at only 1·75, in place of twice the shearing area of rivets in single shear. Machine riveting is generally stronger than hand riveting, and experiments by Messrs. Greig and Eyth indicate an improvement in rivets made by the quickly applied stroke of a steam riveter, or a hydraulic riveter with a quick moving accumulator, over that done by one with a slow moving accumulator.§ They found, however, " that the plate, especially if soft, is much less injured by hydraulic riveting, and that this method has, therefore, a decided advantage where the plate is the weaker part; but that the rivet itself is stronger when put in by the steam riveter, owing probably to the greater compact-

---

* Trans. Inst. Nav. Arch., 1882, p. 115.
† Trans. Am. Soc. C. E., Vol. X., p. 139.
‡ Proc. Inst. C. E., Vol. LXXVII., p. 263.
§ Proc. Inst. M. E., 1879, p. 271.

ness of the rivet material obtained by the sudden shock."[*]
Experiments by Mr. Kirk seem to show that a tighter joint
may be obtained with thick plates if the pressure on the
head of the rivet is sustained for a sensible period, so that the
plates may be pressed together while the rivet is cooling; it
then maintains its contractile grip on the plates without any
risk of their springing apart while the rivet is still hot and
ductile.[†]

3. *Size and Pitch of Iron Rivets—Proportions of Joints—
Boilermakers' Practice—Shipbuilders' Practice.*—The diameter
of iron rivets in boiler work is generally about twice the thickness of the plate when the latter does not exceed $\frac{3}{8}$ inch; after
this the proportion is gradually reduced until the diameter of
the rivet nearly equals the thickness of the plate, the limit of
diameter being about $1\frac{1}{4}$ inches, as this is the largest rivet
that can be conveniently worked in practice.

Tables VII., VIII., and IX., give Boilermakers' proportions for riveting. Table VII. shows the practice of various
manufacturers for riveted iron joints. It has been compiled
by Mr. Tweddell, who says—"With rivets up to 1-inch
diameter it seems to be the universal practice to make the
'margin,' or distance from outside of rivet to edge of plate,
equal to the diameter of rivet. With very large rivets,
$1\frac{1}{16}$, $1\frac{1}{8}$, or $1\frac{1}{4}$-inch diameter, some makers allow rather less
margin—namely, 1, $1\frac{1}{16}$, $1\frac{1}{8}$-in."[‡]

Table VIII. is copied from Seaton's Manual of Marine
Engineering, p. 348, and for Table IX. the author is indebted
to Mr. Aspinall, President of the Institution of Civil Engineers of Ireland.

---

[*] Proc. Inst. M. E., 1879, p. 278.
[†] *Idem*, 1881, p. 329.
[‡] *Idem*, p. 293.

THE STRENGTH AND PROPORTIONS

TABLE VII.—*Boilermakers' Proportions for Lap joints, Iron Plates and Iron Rivets* (Tweddell).

Thickness of plate.	Lap joints, single-riveted.		Lap joints, double-riveted.		
	Diameter of rivet, (nominal).	Pitch of rivets.	Diameter of rivet, (nominal).	Pitch of rivets.	Distance between pitch lines of two transverse rows of rivets.
inch.	inch.	inch.	inch.	inch.	inch.
3/16	3/8	1 1/4	—	—	—
1/4	1/2	1 1/2	—	—	—
5/16	5/8	1 5/8	5/8	2 1/4 to 2 1/2	1 1/4 to 1 3/4
3/8	5/8 to 3/4	1 3/4 to 2	3/4 to 7/8	2 1/2 to 2 11/12	1 7/16 to 2
1/2	3/4 to 7/8	2 to 2 1/4	3/4 to 7/8	2 1/2 to 3 1/8	1 1/2 to 2 1/4
5/8	7/8 to 1	2 1/4 to 2 3/4	7/8 to 1	2 3/4 to 3 3/8	2 to 2 1/4
3/4	1 to 1 1/8	2 1/4 to 2 1/2	1 to 1 1/8	3 to 4 1/8	2 to 2 1/2
7/8	1 to 1 3/8	2 1/2 to 3	1 1/8	3 1/4 to 3 5/8	2 5/8
1	1 to 1 1/4	2 1/2 to 2 3/4	1 1/4 to 1 1/2	3 1/2	2 1/4 to 2 5/8
1 1/8	1 1/4 to 1 3/8	2 5/8 to 3	1 3/8	3 3/4	2 7/8

TABLE VIII.—*Boilermakers' Proportions for Butt Joints, Double Straps, Double-riveted Zigzag; Iron Plates and Iron Rivets* (Seaton).

Thickness of plate.	Diameter of rivets, (nominal).	Pitch of rivets.	Breadth of straps.	Thickness of straps.
inch.	inch.	inch.	inch.	inch.
3/4	7/8	3 1/2	8 7/8	1/2
13/16	15/16	3 3/4	9 3/8	1/2
7/8	1	4	10	9/16
15/16	1 1/16	4 1/4	10 3/8	5/8
1	1 1/8	4 1/2	11 1/4	5/8
1 1/16	1 3/16	4 3/4	11 1/2	11/16
1 1/8	1 1/4	5	12 1/2	3/4
1 3/16	1 5/16	5 1/4	13 1/4	3/4
1 1/4	1 5/16	5 1/16	13	13/16

TABLE IX.—*Locomotive Boiler Proportions for Rivetting* (Aspinall).

Railway Company.	Diameter of boiler, inside.	Thickness of plates.	Single-riveted joints. Form of joint.	Single-riveted joints. Diameter of rivet.	Single-riveted joints. Pitch.	Single-riveted joints. Lap.	Single-riveted joints. Width of butt strips.	Double-riveted joints (zigzag). Form of joint.	Double-riveted joints (zigzag). Diameter of rivet.	Double-riveted joints (zigzag). Pitch.	Double-riveted joints (zigzag). Lap.	Double-riveted joints (zigzag). Width of butt strips.	Remarks.
	ft. in.	in.		in.	in.	in.	in.		in.	in.	in.	in.	
London and North-Western	4 0⅞	$\frac{13}{32}$	Lap for circular seams, double-covered and single-riveted butt for horizontal seams.	¾	2	2¼	5	—	—	—	—	—	Rivets and plates, both steel made by L. & N. W. Ry. Holes punched and annealed. Heads of rivets snap.
Great Western -	4 1½	$\frac{7}{16}$	Lap for circular seams.	¾	1⅝	2½		Double-covered butt for longitudinal seams.	¾	2⅛	—	8½	Rivets and plates, both steel made by Landore Siemens Co. Holes punched. Heads of rivets snap.
London, Brighton, and South Coast	4 5	½	Lap for circular seams.	¾	1⅞	—	—	Butt.	⅞	3⅜	—	9¼	Plates, iron made by Monkbridge Co. Rivets, Yorkshire iron. Holes drilled after bending. Rivets, cup for lap joints and countersunk for butt joints.
Great Northern -	4 2	½	—	—	—	—	—	Lap.	¾	1¾	3½	—	Plates, principally iron, either Lowmoor or Bowling or Cooper. Rivets, Yorkshire iron by same makers as plates. Holes drilled and punched. Heads of rivets snap.
Great Southern and Western	4 0	½	Lap for circular seams.	¾	1⅝	2¼	4½	Double-covered butt for longitudinal seams.	¾	3¾	—	6¾	Plates, steel made by Bolton Iron and Steel Co. Rivets, Lowmoor. Holes drilled $\frac{27}{32}$''. Heads of rivets snap.

Table X.—Lloyd's Proportions for Ship-riveting (Iron).

	ins.	ins.	ins.	ins.	ins.	ins.	ins.	ins.	ins.	ins.	ins.	ins.	ins.	ins.	ins.
Thickness of Plates,	$\frac{5}{16}$	$\frac{6}{16}$	$\frac{6\&7}{16}$ alternately	$\frac{7}{16}$	$\frac{8}{16}$	$\frac{9}{16}$	$\frac{9\&10}{16}$ alternately	$\frac{10}{16}$*	$\frac{11}{16}$	$\frac{12}{16}$	$\frac{12\&13}{16}$ alternately	$\frac{13}{16}$	$\frac{14}{16}$	$\frac{15}{16}$	$\frac{16}{16}$
Diameter of Rivets,	$\frac{5}{8}$	$\frac{5}{8}$	$\frac{3}{4}$	$\frac{3}{4}$	$\frac{3}{4}$	$\frac{3}{4}$	$\frac{7}{8}$	$\frac{7}{8}$	$\frac{7}{8}$	$\frac{7}{8}$	1	1	1	$1\frac{1}{8}$	$1\frac{1}{8}$
Breadth of Treble-riveted Straps,	—	—	—	—	$14\frac{1}{4}$	$14\frac{1}{4}$	$14\frac{1}{4}$	$16\frac{3}{4}$	$16\frac{3}{4}$	$16\frac{3}{4}$	19	19	19	$21\frac{1}{4}$	$21\frac{1}{4}$
" Double "	—	8	$9\frac{3}{4}$	$9\frac{3}{4}$	$9\frac{3}{4}$	$9\frac{3}{4}$	$9\frac{3}{4}$	$11\frac{1}{4}$	$11\frac{1}{4}$	$11\frac{1}{4}$	—	—	—	—	—
" Double-riveted Laps,	8	$3\frac{3}{4}$	$4\frac{1}{4}$	$4\frac{1}{2}$	$4\frac{1}{2}$	$4\frac{1}{2}$	$4\frac{1}{2}$	$5\frac{1}{4}$	$5\frac{1}{4}$	$5\frac{1}{4}$	6	6	6	$6\frac{3}{4}$	$6\frac{3}{4}$
" Single-riveted "	$3\frac{3}{4}$	$2\frac{1}{4}$	$2\frac{1}{4}$	$2\frac{1}{2}$	$2\frac{1}{2}$	—	—	—	—	—	—	—	—	—	—
Maximum spacing of Rivets from centre to centre, in Butts of outside plating,	$2\frac{1}{4}$	$2\frac{1}{2}$	3	3	3	3	3	$3\frac{1}{2}$	$3\frac{1}{2}$	$3\frac{1}{2}$	4	4	4	$4\frac{1}{2}$	$4\frac{1}{2}$
" Edges "	$2\frac{3}{4}$	$2\frac{3}{4}$	$3\frac{3}{8}$	$3\frac{3}{8}$	$3\frac{3}{8}$	$3\frac{3}{8}$	$3\frac{3}{8}$	4	4	4	$4\frac{1}{2}$	$4\frac{1}{2}$	$4\frac{1}{2}$	5	5
" Frames "	5	5	6	6	6	6	6	7	7	7	8	8	8	—	—

* Where stringer and tie plates are $\frac{10}{16}$ of an inch thick, they should be secured to the beams with $\frac{3}{4}$ of an inch rivets. Where the fore and aft flange of the frame does not exceed 3 inches, the rivets attaching the outside plating thereto should not exceed $\frac{7}{8}$ of an inch in diameter, and where it is $3\frac{1}{2}$ inches wide, they should not exceed 1 inch in diameter. Rivets to be $\frac{3}{4}$ of an inch larger in diameter in the stem, stern-post, and keel; but in no case need these exceed $1\frac{1}{4}$ inches in diameter.

M. Antonie gives the following empirical formula for the diameter of iron rivets in shipbuilders' practice.*

$$d = 1 \cdot 1 \sqrt{t} \qquad (1)$$

where d = diameter of rivet,
t = thickness of plate.

Tables X., XI., and XIII., give Shipbuilders' proportions for riveting, copied from the rules of the Underwriters of Lloyd's, Liverpool, and Bureau Veritas.

Lloyd's Rules.—"The butts of outside plating to be chain-riveted. All double and treble riveting, except in the keel, stem, and stern-post, is recommended to be chain riveting. In chain-riveted butts, a space equal to twice the diameter of the rivet to be between each row; where treble riveting is adopted, a space equal to twice the diameter of the rivet to be between each row, with half the number of rivets in the back row. The overlaps of plating, where chain riveting is adopted, are not to be less than six times the diameter of the rivets; and, where single riveting is admitted, to be not less than three and a half times the diameter of the rivets. The rivets are not to be nearer to the butts or edges of the plating, butt straps, or of any angle iron, than a space equal to their own diameter; and, in edge rivet-ing, the space between any two consecutive rows of rivets must not be less than once and a half their diameter. The rivet holes to be regularly and equally spaced and care-fully punched from the faying surfaces opposite each other in the adjoining parts, laps, lining pieces, butt straps, and frames; and countersinking to extend through the whole thickness of the plate or angle bar. They are to be spaced not more than four diameters of the rivet apart from centre to centre in the butts of the plating, and not more than from four to four and a half diameters apart in the edges of the

* Proc. Inst. C. E., Vol. XXII., p. 200.

plating and at other parts, excepting in the keel, stem, and stern-post, where they may be five diameters, and through the frames and outside plating, and in reversed angle iron on frames, where they may be eight diameters apart from centre to centre. The rivets in the flanges of the gunwale angle irons to be spaced not more than four and a half diameters apart from centre to centre, and those connecting iron decks and stringer plates to the beams to be spaced from seven to eight diameters apart. The rivets are to be of the best quality, and to be in diameter as per Table X., and to be increased in size under their heads to fill the rivet holes. When riveted up, the rivets are completely to fill the holes, their heads are to be 'laid up,' and their points or outer ends are not to be below the surface of the plating."

TABLE XI.—*Liverpool Proportions for Ship Riveting (Iron).*

Thickness of Plates, in sixteenths of an inch, - -	5	6	7	8	9	10	11	12	13	14	15	16
Diameter of Rivets, in sixteenths of an inch, - -	10	10	12	12	13	14	14	16	16	16	17	18
Breadth of Lap in Seams, in inches.												
Single riveting, -	2½	2½	3	3	—	—	—	—	—	—	—	—
Double riveting, -	3¾	3¾	4½	4½	5	5¼	5¼	6	6	6	6½	6¾
Breadth of Butt Strip, in inches.												
Double riveting, -	8	8	10	10	10¾	11½	11½	13	13	13	13¾	14½
Treble riveting, -	12	12	14½	14½	15¾	17	17	19	19	19	19¾	21½
Quadruple riveting,	16	16	19	19	20¾	22½	22½	25	25	25	26¾	28½

Liverpool Rules.—"Rivets to be 4 diameters apart from centre to centre, longitudinally in seams and vertically in butts, except in the butts where treble or quadruple riveting is required, when the rivets in the row farthest from the butt may be spaced eight diameters apart, centre to centre. Rivets in framing to be eight times their diameter apart from centre to centre, and to be of the size required by table. Rivets in shell plating, in deck ties and stringers, in centre plate, and in flat keels and keelsons, are to have their necks bevilled under their heads, so as to fill the countersink made in punching, and their heads should not be thicker than two-thirds of the diameter of the rivet, and it is recommended that the necks of all rivets be bevilled under the heads. The rivet holes in outside plating, deck stringers and ties, gunwale and gutter angle-irons and iron decks, to be carefully countersunk quite through the plate or bar. The face, or largest diameter of the countersink, should not be less than one and a half times the diameter of the rivet. All butts of plates to be chain riveted. All butt and seam holes must be punched from the surfaces which are placed together, so that the taper of the holes shall be in opposite directions. The holes to be punched fair, and opposite each other; unfair holes will render the piece of work badly punched liable to rejection. Where holes cannot be truly punched they must be drilled through fair. Butts to be closely fitted, either by planing or jumping; when jumped, the ridge formed by jumping to be chipped off the inside, in order that the butt strips may fit closely. The ridge outside to be hammered into the joint. Butt strips to have grain of iron in the same direction as, and to be of not less thickness than, the plates which they connect. When double butt strips are fitted, they may each be two-thirds the thickness of the plates united by them. Treble riveted strips to be at least ten per cent. thicker than the plates they connect. Treble riveted strips on sheer strakes of and over twelve-sixteenths thick, to be at least fifteen per cent. thicker than the plates they connect. Quadruple riveted strips on sheer

B

strakes of and over fourteen-sixteenths thick, to be at least twenty per cent. thicker than the plates they connect."

TABLE XII.—*Bureau Veritas, Proportions for Ship Riveting* (*Iron*).

Thickness of Plates.	Diameter of Rivets		Breadth of Laps for			Width of Butt Straps and Thickness for	
	for Plates and Angle Irons.	for bar Keels, Stem, and Sternpost.	Single Riveting.	Double Zigzag Riveting.	Double Chain Riveting.	Double Riveting.	Treble Riveting.
16ths.	16ths.	16ths.	In.	In.	In.	In. 16ths.	In. 16ths.
4	8	12	$1\frac{3}{4}$	$2\frac{3}{4}$	3	$6\frac{1}{4} \times 4$	$9\frac{1}{4} \times 5$
5	9	13	2	$3\frac{1}{8}$	$3\frac{5}{8}$	7×5	$10\frac{3}{8} \times 6$
6	10	14	$2\frac{1}{4}$	$3\frac{5}{8}$	$3\frac{3}{4}$	$7\frac{1}{2} \times 6$	$11\frac{1}{4} \times 7$
7	11	15	$2\frac{5}{8}$	$3\frac{7}{8}$	$4\frac{1}{8}$	$8\frac{1}{4} \times 7$	$12\frac{3}{8} \times 8$
8	12	16	$2\frac{3}{4}$	4	$4\frac{1}{2}$	9×8	$13\frac{1}{2} \times 9$
9	13	17	—	$4\frac{7}{16}$	$4\frac{7}{8}$	$9\frac{3}{4} \times 9$	$14\frac{5}{8} \times 10$
10	13	17	—	$4\frac{7}{16}$	$4\frac{7}{8}$	$9\frac{3}{4} \times 10$	$14\frac{5}{8} \times 12$
11	14	18	—	$4\frac{3}{4}$	$5\frac{1}{4}$	$10\frac{1}{4} \times 11$	$15\frac{3}{8} \times 13$
12	14	18	—	$4\frac{3}{4}$	$5\frac{1}{4}$	$10\frac{3}{8} \times 12$	16×14
13	15	19	—	5	$5\frac{5}{8}$	$11\frac{1}{4} \times 13$	$16\frac{7}{8} \times 15$
14	16	20	—	$5\frac{7}{16}$	6	$12\frac{1}{4} \times 14$	$18\frac{1}{4} \times 16$
15	17	20	—	$5\frac{3}{4}$	$6\frac{3}{8}$	—	$19\frac{1}{4} \times 17$
16	18	21	—	$6\frac{1}{8}$	$6\frac{3}{4}$	—	$20\frac{1}{4} \times 18$

Bureau Veritas, Rules.—"The rivets to be of the best description of fibrous iron, and in size according to the table. The holes to be regularly pitched, and carefully punched opposite each other from the joint surfaces. The countersinking of the outside plating and stringers to extend through two-thirds of the thickness of the plate; the holes not to be nearer to the edge of any plate or angle bar than the diameter of the rivet. Rivets in the outside plating to be laid up round the heads, to fill the holes and countersink, and be finished flush on the outside. The keel rivets may be left full or convex. The longitudinal spacing of rivets in seams of plating, bulkheads, and for all water-tight parts not to exceed four diameters from centre to centre for single riveting, and four and a half diameters for double riveting. The riveting of butts for the two first rows to be four diameters from centre to centre, and eight diameters for the third row. The spacing of rivets in keel, stem, and sternpost not to exceed five diameters from centre to centre. In chain riveting the distance from centre to centre between the rows to be equal to three times, and in zigzag riveting to be equal to twice the diameter of the rivet. The diameter of rivets and system of riveting to be determined by the thickest plate."

The rivets in girder work are generally the same size as those adopted by boiler makers, but, as girders do not require caulking like a boiler, the pitch, or distance of rivets from centre to centre, is much greater, and usually varies from 3 to 5 inches, but it should not exceed ten to twelve times the thickness of a single plate, as otherwise damp may get between the plates and cause rust, which in time swells and bursts them asunder. In girder work, also, the tension joints are generally triple or quadruple riveted, to enable sufficient rivet area to be obtained without weakening the plates by numerous transverse perforations ; and Mr. Milton has lately drawn attention to a matter which, though previously known, was seldom acted on—namely, that the shearing strength of rivets in triple and quadruple riveting is theoretically less than in single or double riveting, in consequence of the unequal stress and unequal stretching of the plates between the several rows of rivets.* This brings an undue stress on the outer rows of rivets, and it seems therefore desirable to provide an excess of rivet area, say 10 per cent., in triple or quadruple riveted

* Trans. Inst. Nav. Arch., 1885.

joints to compensate for this unequal distribution of stress among the rivets. This remark probably applies with considerable force to joints in the flanges of girders when they are formed with piles of plates, but is of less importance in ordinary cases, for in Kirkaldy's experiments for the Board of Trade on triple-riveted lap joints with steel plates and steel rivets, the shearing strength of the latter does not appear to have been at all diminished by the unequal stress in the plates referred to by Mr. Milton.*

Mr. Moberly informs me that the practice at the works of Messrs. Easton and Anderson of Erith is to make the punch about $\frac{1}{16}$ inch (that is, from 5 to 12 per cent.) larger in diameter than the rivet (the larger percentage being used with the smaller rivets), and to make the die larger than the punch at the rate of $\frac{1}{8}$ inch for 1 inch thickness of plate. Hutchinson also says the difference between the punch and the die may be taken as not much less than $\frac{1}{8}$ of an inch to the inch in thickness;† thus, in punching for a 1-inch rivet in a plate 1 inch thick, the punch will be $1\frac{1}{16}$ inch and the die will be $1\frac{3}{16}$ inch in diameter, and for the same rivet in a $\frac{1}{2}$ inch plate the punch will be $1\frac{1}{16}$ inch and the die will be $1\frac{1}{8}$ inch in diameter. This clearance of the die outside the punch equals 12·5 per cent. of the thickness of the plate, or say 6 per cent. of the nominal diameter of the rivet for plates not exceeding $\frac{5}{8}$ inch in thickness, but increasing up to about 10 per cent. of the nominal diameter of the rivet when the latter reaches the larger sizes above 1 inch in diameter. As the punched hole is conical, its diameter at the top being equal to, or slightly larger than that of the punch, and at the bottom equal to the diameter of the die, the mean diameter of the punched hole is found by adding the nominal diameter of the rivet to half the clearances of the punch

---

* Merchant Shipping Experiments, pp. 83, 85.
† Hutchinson's Girdermaking, p. 53.

and die added together. Taking a ¾ inch rivet with ⅜ inch plates as an average example, we find that the clearance of the punch is $\tfrac{4}{64}$ inch, or nearly 8 per cent. of the nominal size of the rivet; and the hole in the die is $\tfrac{3}{64}$ inch larger again than the punch, that is, its clearance is $\tfrac{7}{64}$ inch, or 14 per cent. of the nominal size of the rivet. The mean clearance of the punched hole for a ¾ inch rivet is therefore about 11 per cent. of the nominal size of the rivet, or even more if the die be somewhat worn by use. In boiler work the plates are put together with the larger, or countersunk, ends of the holes outwards, and the shearing diameter of the finished rivet is equal to, or a little greater than that of the punch, which in the case of our ¾ inch rivet is 8 per cent. larger than the nominal size of the rivet. The following table will illustrate this example more clearly:—

	Inch.		Inch.		Inch.	
Nominal size of ¾ inch rivet,	$1\tfrac{2}{16}$	=	$\tfrac{48}{64}$	=	0·75	
Diameter of punch, ...	$1\tfrac{3}{16}$	=	$\tfrac{52}{64}$	=	0·8125	
Diameter of hole in die, ...			$\tfrac{55}{64}$	=	0·8594	
Diameter of conical hole } at top					0·8125 }	mean 0·836
produced by punching, } at bottom					0·8594 }	

Wilson says that " the allowance made in the length of a rivet for forming the head should be about 1¼ times the diameter for snap and conical heads, and about equal to the diameter for countersunk heads. In machine riveting the length requires to be ⅛ to ¼ inch more than the above."* This affords an additional reason for machine riveting being somewhat stronger than hand riveting, inasmuch as more metal is squeezed into the rivet by the machine than by hand.

4. *Tensile Strength of Perforated Plates.*—The theoretic percentage of the strength of a perforated plate at any joint, as compared with that of the solid plate, is as follows:—

* Wilson on Steam Boilers, p. 55.

$$\text{Theoretic percentage} = \frac{p-d}{p} \times 100 \qquad (2)$$

where p = pitch of rivets (in inches),

d = diameter of rivets (in inches).

This theoretic percentage is, however, rarely or never obtained in practice with iron plates, as will appear below. Nearly all experimenters on the subject agree that punching generally reduces the tenacity of iron and steel plates to a greater degree than the area of the metal punched out, and a close examination of the border of each hole shows that it has been subject to a certain degree of violence, which in most cases has reduced the ductility of the metal, and made it locally crystalline in fracture, and, as some suppose, caused incipient cracks round the edge of the hole; but this latter seems doubtful, as Mr. Beck-Gerhard, of the Poutiloff works at St. Petersburg, instituted "an investigation as to whether there was any foundation for the very generally received opinion that the edges of a punched hole on the die side are injured by a ring of minute incipient cracks. For this purpose a large number of specimens, 5 inches by 3 inches by ½ inch of all kinds of steel were prepared. The edges were planed, the surfaces polished, holes were pierced in various ways, and the metal surrounding them was carefully examined with a microscope, but no trace whatever of cracks could be found, though the nature of the steel ranged from 0·1 to 0·6 per cent. of carbon."[*] Owing to its hardness and inability to stretch, this annulus of strained material round the punched holes, when the specimen is under tension, takes a higher proportion of the stress than the other more yielding parts, and hence it reaches the breaking-point sooner—that is, the punched plate breaks in detail:—first the annulus of hard metal gives way, and afterwards the more ductile portion between the holes. Rimering, or boring out, a zone of metal

[*] Proc. Inst. C. E., Vol. LXXVII., p. 450.

⅛th inch wide round the punched hole removes the annulus of strained material and neutralises the effect of punching. Countersinking has also the same effect,* and annealing after punching also restores the ductility and original tenacity of the metal, but it is seldom or never considered necessary to anneal iron plates, though this is often done with thick steel ones. In numerous experiments on the subject† the loss of tenacity in iron plates from punching varied from 5 to 23 per cent. of the original strength of the solid plate, but the percentage in any particular case will doubtless depend—1°. On the diameter of the holes. 2°. On their pitch. 3°. On the width of the strip punched, for wide plates are apparently less injured than narrow strips, perhaps because it is difficult to centre the holes accurately in the latter. 4°. On the condition of the punching tool—*i. e.*, the sharpness of its cutting edges, and the maintenance of the proper proportion of size between the punch and the die. 5°. On the quality and thickness of the metal, hard iron generally suffering more than ductile iron, and thin plates less than thick ones.‡ Probably, the most accurate method for making an allowance for the injurious effect of punching would be to add a certain percentage, say one-tenth to one-fifth of its diameter to each hole, when calculating the effective net area of a punched plate. Mr. White states that "in making their estimate for riveted work in the Royal Navy, they were accustomed to allow 4 tons off the very best iron for punching—4 tons off 22 tons."§ Civil Engineers, however, rarely make any similar allow-

---

* The Admiralty practice in countersinking steel plates is to punch the hole ¼-inch less than the smaller end of the countersink, which is carried through the full thickness of the plate, and thus removes all the strained material.—Wildish, Trans. Inst. Nav. Arch., 1885.

† Proc. Inst. M. E., 1881, p. 323.

‡ In Mr. Sharp's experiments on steel plates, the reduction of tenacity due to punching was greater the harder the material of the plates, but annealing restored them to their original strength.—Proc. Inst. M. E., 1881, p. 307.

§ Proc. Inst. C. E., Vol. LXIX., p. 36.

ance in riveted girder work, perhaps because the rivet holes are generally pitched farther apart in girders than in ships. Drilling does not strain or damage the metal surrounding the hole, and it is, therefore, to be preferred for first-class work. Some experiments seem to show that the unit-strength of a drilled iron plate is greater than that of the solid plate, owing, it is supposed, to the flow of the metal between the holes, and consequent contraction of area during the last stages of testing, being restrained by the adjoining solid parts—or, in other words, the drilled plate does not contract sideways and stretch lengthwise in the line of transverse holes as much as a solid plate would from the same unit-stress, and the contraction of area at the line of fracture is thus less in the perforated than in the solid plate. With punched plates, however, as already explained, the action of the punch causes initial stresses in the annulus of metal around each hole and thus alters its power of stretching, so that at the crisis of fracture this strained part gets an excessive proportion of the load and breaks before the other parts of the plate are strained to their full capacity. This increase of strength from drilling, however, is so extremely doubtful with iron that it cannot be depended on for increasing the tenacity of a drilled plate. With mild steel, however, it is more important. Mr. Longridge states that in a long series of experiments on perforated iron plates, whether the holes were punched or drilled, there was an absolute loss of unit-strength from perforation. In some few of the drilled plates there was a gain, but only a slight one.* In the author's experiments on drilled iron plates, though some of them showed a gain from perforation, the majority showed an absolute loss.†

5. *Lap of Plates and Pitch of Rivets.*—A joint may fail

---

* Proc. Inst. M. E., 1881, p. 267.
† Trans. Roy. I. A., 1875, p. 454.

from the holes being too close to the edge of the plate, in which case the rivet splits or bursts open the margin of the plate in front of itself. Experiments indicate that the margin of single-riveted lap joints need not exceed one diameter of the rivet (§ 3)—that is, the lap of single-riveted lap joints need not exceed three times the nominal diameter of the rivet for plates; with bars, however, the lap should not be less than from $3\frac{1}{2}$ to 4 diameters of the rivet, as the end of a narrow bar is evidently more liable to split open than the edge of a wide plate, where the adjoining parts hold the margin up to its work.* In boiler work the lap is generally three nominal diameters of the rivet for single-riveted lap joints and about five nominal diameters for double-riveted lap joints; if the lap much exceeds five diameters the plates are not close along their edges, and it is difficult to make the seam steam-tight by caulking. In shipbuilding, however, the lap is somewhat greater than in boilerwork (*see* Tables X., XI., XII.), and in girder work, where the edges of the plate are often roughly shorn, the margin, or distance between the rivet holes and edge of the plates, is seldom less than $1\frac{1}{4}$ times the diameter of the rivet. Browne observes that, "as the effect of punching is to weaken the plate to some distance all round the punched hole, the result will be that in the space between any two successive holes in the straight line of rivets the plate is weakened twice the distance that the punching affects, but in the zigzag line between the same two holes the plate is weakened to the extent of four times the same distance," and he recommends the longitudinal distance between the pitch lines in zigzag riveting to be two-thirds of the transverse pitch. "In chain riveting, however, the rivets in the second row, being opposite to those in the first row, are in the same position with respect to the first row as the rivets in a single-riveted joint to the edge of the

---

* Trans. Roy. I. A., Vol. XXV., p. 453.

lap. Hence, by the same rule as before, the distance between the rivet holes in the two rows will be one diameter, making the distance between the pitch lines 2 diameters; but as the plate between the holes will be injured at both sides by punching, it will be safer to make the distance $2\frac{1}{2}$ diameters between the pitch lines."* The rules thus obtained by Browne for double-riveted lap joints with punched holes are briefly as follows:—

    Diameter of rivet = 2 times thickness of plate.

    Pitch - - = $4\frac{1}{2}$ diameters of rivet.†

    Lap - - $\begin{cases} = 5\frac{1}{2} \text{ diameters in chain riveting.} \\ = 6 \text{ diameters in zigzag riveting.} \end{cases}$

Wilson says that the width of the strap for double-riveted butt joints in boiler work should be at least nine times the diameter of the rivet, and may with thick plates be ten times.‡ In shipbuilding the width of the strap is much greater (*see* Tables X., XI., and XII.).

6. *Lap joints, effect of bending.*—Single-riveted lap joints bend considerably under severe stress, and the plates are then

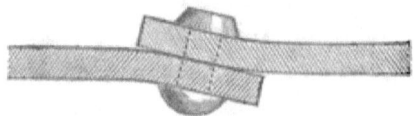

liable to tear, not only from bending along the line of holes, but because they are subject to an unequally distributed crushing stress, in consequence of the rivets pressing more on the inner than on the outer edge of each plate. The unit-strength of the plate is thus much less in single-riveted lap joints than in perfectly straight joints, such as double-covered butt joints, or joints where they are bent but slightly, such as double and triple-riveted lap joints. Indeed, Fairbairn concludes

---

* Proc. Inst. M. E., 1872, pp. 62, 63, 64.

† This pitch is rather large for steam-tightness. The pitch in double-riveted lap joints in boiler work seldom exceeds $3\frac{1}{2}$, or at most 4, diameters of the rivet.

‡ Wilson on Steam Boilers, p. 85.

from his experiments that the unit-strength of the iron between the holes of double-riveted lap joints is equal to that of the solid plate.* It will also be observed that the rivets in the bent joints are in tension as well as in shear, and if their heads are too small they are sometimes pulled off in testing the joint.

TABLE XIII.—*Experiments showing the Reduction of Unit-strength of the Plates in Single-riveted Lap joints compared with their original Unit-strength before Perforation.*

	Ratio of tensile unit-strength of net area of plate in the joint to the unit-strength of the original solid plate.	Experimenter.	Remarks.
	per cent.		
Joints with drilled holes.	71	Stoney	$\frac{3}{8}$ in. plates and five $\frac{3}{4}$ in. rivets, hand riveted.
	95	,,	Do. do.
	101	,,	Do. do.
	76	,,	Do. do.
	100·5	,,	Do. do.
	86	,,	Do. do.
	99	,,	Do. do.
	84	Greig & Eyth	$\frac{3}{8}$ in. plates and four $\frac{5}{8}$ in. rivets, steam riveted.
	97	,,	$\frac{3}{8}$ in. plates and four $\frac{3}{4}$ in. rivets.
Joints with punched holes.	79	Stoney	$\frac{3}{8}$ in. plates and five $\frac{3}{4}$ in. rivets, hand riveted.
	65	,,	Do. do.
	96	,,	Do. do.
	92	,,	Do. do.
	75	,,	Do. do.
	78	,,	Do. do.
	76	,,	Do. do.
	75·5	Greig & Eyth	$\frac{3}{8}$ in. plates and four $\frac{5}{8}$ in. rivets, steam riveted.

(Average for drilled holes, 89·94 per cent. Average for punched holes, 79·6 per cent.)

* Fairbairn's Useful Information for Engineers, 1st Series, p. 283.

7. *Single-riveted Lap joints—Reduction of Unit-strength of Plate—Efficiency of Joints.*—Table XIII. contains the results of 14 experiments by the author, and 3 by Messrs. Greig and Eyth, on single-riveted lap-joints of fairly good proportions.*

Reducing these results to a simple standard, we have—

Tensile unit-strength of the solid plate, - - 100
" of the plate remaining between the holes in single-riveted lap joint with drilled holes, - 90
" ditto with punched holes, - 80

We thus see that from the mechanical treatment—*i.e.*, drilling or punching—and from the joints bending under severe stress, the iron which remains between the holes in a single-riveted lap joint with $\frac{3}{8}$-inch plates of fair proportions loses 10 per cent. of its original unit-strength when the holes are drilled, and 20 per cent. when they are punched. This result agrees closely with Fairbairn's conclusion, from his experiments on thinner punched plates, that the unit-strength of the iron between the holes of single-riveted joints is 79 per cent. of that of the original plate.† To obtain the actual strength of the joint we must make a farther reduction for the metal cut out by the holes, which reduces the strength of the joint to little more than one-half of the foregoing percentages, as shown by the following abstract, which has been derived from experiments by the author and others on well-proportioned single-riveted lap joints in which the tensile strength of the plates and the shearing strength of the rivets were pretty fairly balanced:‡—

---
* Trans. Roy. I. A., Vol. XXV., p. 451, and Proc. Inst. M. E., 1879, p. 297.
† Useful Information for Engineers, 1st Series, p. 283.
‡ Proc. Inst. M. E., 1881, pp. 340, 341; Trans. Roy. I. A., Vol. XXV., p. 451.

TABLE XIV.—*Experiments showing the Efficiency of Single-riveted Lap joints in percentages of the strength of the original solid plate.*

\multicolumn{3}{c}{PUNCHED HOLES.}	\multicolumn{3}{c}{DRILLED HOLES.}
Experimenter.	Efficiency of joint, per cent.	Broken by rivets shearing or by plates tearing.	Experimenter.	Efficiency of joint, per cent.	Broken by rivets shearing or by plates tearing.
Fairbairn	46	Plates torn	Stoney, ⅜-in. plates.	45	Plates torn
,,	46	,,	,,	55	,,
,,	44	,,	,,	50	,,
,,	44	Rivets sheared	,,	44	,,
Kirkaldy	54	Plates torn	,,	50	,,
Stoney, ⅜-in. plates.	50	,,	,,	50	,,
,,	44	,,	,,	44	,,
,,	47	,,	,,	46	Rivets sheared.
,,	45	,,	,,	50	,,
,,	45	,,	Greig & Eyth	50·4	Plates torn.
,,	47	,,	,,	46·5	,,
,,	37	,,	,,	50·8	,,
,,	38	Rivets sheared	,,	53·9	Rivets sheared.
,,	46	,,	,,	57·6	,,
Greig & Eyth	40.6	Plates torn	,,	—	—
Mean, 44·9			Mean, 49·5		

The efficiency of a joint is the ratio of the strength of the joint to the strength of an equal width of solid plate, and from the foregoing table we may infer that the average efficiency of single-riveted lap joints, with ⅜-inch iron plates and with drilled holes, does not exceed 50 per cent. of the solid plate, and that with punched holes it does not exceed 45 per cent. of the solid plate. This view is strengthened by the following table, showing the calculated efficiency of single-riveted lap joints of ordinary boilermakers' proportions.

TABLE XV.—*Calculated Efficiency of Single-riveted Lap joints with Punched Holes, Boilermakers' proportions.*

	A	B	C	D	E	F	G	H	I	J	K	L
	Thickness of Plate.	Nominal diameter of Rivets, i.e., their diameter before heating.	Pitch of Rivets.	Diameter of Punch, $= B + \frac{1}{16}$ inch.	Diameter of Die, $= D + 12.5$ per cent. of A.	Mean Diameter of Hole, $= \frac{D+E}{2}$	Shearing area of one rivet, if plates are properly put together, $=$ area of hole at smaller end, $=$ area of punch $= \frac{\pi D^2}{4}$	Net area of plate between two holes, $= A(C-F)$.	Shearing strength of one rivet, $= G \times 19$ tons.	Tensile strength of plate between two holes, $= H \times 18$ tons.	Tensile strength of solid plate for width of one pitch, $= A \times C \times 22.5$ tons.	Efficiency of joint, $=$ per cent-age of I or J (whichever is lowest) of K.
	inches.	inches.	inches.	inches.	inches.	inches.	sq. ins.	sq. ins.	tons.	tons.	tons.	per cent.
	$\frac{3}{16}=.1875$	$\frac{3}{8}=.375$	$1\frac{1}{4}=1.25$	$\frac{7}{16}=.4375$	.4609	.4492	.1503	.15	2.855	2.70	5.27	51
	$\frac{1}{4}=.25$	$\frac{1}{2}=.50$	$1\frac{1}{2}=1.5$	$\frac{9}{16}=.5625$	.5937	.5781	.2485	.23	4.721	4.14	8.44	49
	$\frac{5}{16}=.3125$	$\frac{5}{8}=.625$	$1\frac{5}{8}=1.625$	$\frac{11}{16}=.6875$	.7265	.7070	.3712	.2863	7.052	5.162	11.425	45.2
	$\frac{3}{8}=.375$	$\frac{3}{4}=.75$	$1\frac{7}{8}=1.875$	$\frac{13}{16}=.8125$	.8593	.8359	.5185	.3896	9.851	7.012	15.82	44.3
	$\frac{1}{2}=.5$	$\frac{13}{16}=.8125$	$2\frac{1}{8}=2.125$	$\frac{7}{8}=.875$	.9375	.9062	.6013	.6094	11.424	10.97	23.91	45.9
	$\frac{5}{8}=.625$	$\frac{7}{8}=.875$	$2\frac{1}{4}=2.25$	$\frac{15}{16}=.9375$	1.0156	.9765	.6903	.796	13.115	14.33	31.64	41.4
	$\frac{3}{4}=.75$	$1=1.0$	$2\frac{1}{2}=2.50$	$1\frac{1}{16}=1.0625$	1.1562	1.1093	.8864	1.043	16.842	18.77	42.19	39.9
	$\frac{7}{8}=.875$	$1\frac{1}{8}=1.125$	$2\frac{1}{2}=2.50$	$1\frac{3}{16}=1.1875$	1.2969	1.2422	1.1075	1.1	21.042	19.8	49.22	40.2
	$1=1.00$	$1\frac{1}{8}=1.125$	$2\frac{1}{2}=2.50$	$1\frac{3}{16}=1.1875$	1.3125	1.25	1.1075	1.25	21.042	22.5	56.25	37.4

The table has been calculated from the following data, which are based on the experiments and deductions already described in this paper:—

1. The holes are punched.
2. The shearing strength of rivets = 19 tons per sq. inch.
3. The tensile strength of the solid plate = 22·5 tons per square inch.
4. The tensile strength of the plate between the holes, in consequence of punching and bending, in a single-riveted lap joint = 80 per cent. of the tensile strength of the solid plate, = 18 tons per square inch.

In the experiments in Table XIV. on single-riveted lap joints the difference of strength in favour of drilled over punched work is about 10 per cent. of the latter; but experiments are still wanting to show what is the corresponding difference in double-riveted lap joints or double-covered butt joints. In these latter the relative advantage derived from drilling is probably less than in single-riveted lap joints, in consequence of the pitch of the rivets being wider.

8. *Double-riveted Lap joints.*—The unit-strength of the plate in a double-riveted is considerably higher than in a single-riveted lap joint, in consequence of there being fewer perforations in each transverse row of rivets, and also in consequence of the double rows keeping the joint from bending through so great an angle as the single-riveted joint, and the average efficiency of double-riveted lap joints, with iron plates not exceeding $\frac{5}{8}$ inch in thickness and designed for strength (not for steam-tightness), is probably about 60 per cent. of that of the solid plate, as shown by the experiments in Table XVI.:—

TABLE XVI.—*Experiments on Efficiency of Double-riveted Lap joints.*

Mode of riveting.	Punched or Drilled.	Efficiency of joint—i.e., ratio of strength of joint to that of solid plate of same width.	Remarks.	Experimenter.	Authority.
		per cent.			
—	Punched	58 to 60	—	Fairbairn	Unwin. Proc. I. M. E., 1881, p. 345.
—	—	60·3	Plates mostly ½″, and rivets ⅝″ to ⅞″. Mean of 10 experiments.	Kirkaldy	Longridge. Engineer, Feb., 1877.
Machine	—	53·8	—	Kirkaldy	Fletcher. Proc. I. M. E., 1881, p. 275.
Machine	—	64·5	—	Kirkaldy	Moberly. (Communicated.)
Machine	Punched	60	Mean of 3 experiments.	—	Unwin. Proc. I. M. E., 1881, pp. 345, 346.
Machine	Punched	62·3	—	Kirkaldy	Unwin. Proc. I. M. E., 1881, p. 345.
—	Drilled	61·9	Plates ⅜″, and rivets ⅝″ and ¾″. Mean of 2 experiments.	—	Greig & Eyth. Proc. I. M. E., 1879, p. 297.

The efficiency of double-riveted lap joints with thick iron plates, however, probably does not exceed 55 per cent. of the strength of the solid plate.

9. *Butt joints—Crushing Pressure of Rivets.*—Single-covered butt joints, with covers of the same thickness as the plates, are merely modifications of lap joints, and the few experiments on the subject indicate that their efficiency is nearly, but not quite, the same as that of similarly riveted lap joints. If, however, the cover of a single-covered butt joint be made a

good deal thicker than the plate, the bending of the joint will be restrained by the stiffness of the cover, and the strength of the joint will be somewhat increased. The Board of Trade rules for marine boilers specify that, "when single butt-straps are used and the rivet holes in them punched, they must be one-eighth thicker than the plates they cover," and that when double butt-straps are used they should be at least ⅝ths the thickness of the plates they cover. Professor Unwin states that in some experiments butt-straps of equal thickness with the plates proved weaker than the latter, and he recommends that single butt-strips should be made 1½ times the thickness of the plate and double butt-straps, each ⅝ths of the thickness of the plate.* Mr. T. Aveling, who has had much experience in boiler-making, states that when the strap for butt joints with single covers was made 25 per cent. thicker than the boiler plate itself, that would be found to be the best method of forming the longitudinal seams, as well as of coupling the rings of the boiler together.† In double-covered butt joints the rivets are in double shear, and the bearing pressure of each rivet against the central plate is twice as great as in a lap joint, provided the shearing unit-stress of the rivet be the same in both cases. In lap joints this bearing pressure rarely exceeds 30 tons per square inch when the joint is tested to fracture—that is, assuming that the pressure is uniformly distributed over the bearing surface, which in reality is not the case on account of the bending of the joint (see § 6), and there seems some reason for supposing that when the bearing pressure of a rivet in single shear much exceeds this, the tenacity of the plate is reduced.‡ There is also reason for supposing that when the bearing pressure of a rivet in double shear—that is, its bear-

* Unwin's Machine design, p. 87.
† Proc. Inst. M. E., 1879, p. 315.
‡ Proc. Inst. M. E., 1881, p. 333. Trans. Roy. I. A., Vol. XXV., p. 453.

ing pressure against the middle plate of a double-covered butt joint, exceeds 40 tons at the crisis of fracture, the tenacity of the plate is somewhat reduced, and Professor Kennedy concludes from his experiments that the shearing strength of rivets is also reduced by severe bearing pressure. However this may be, the efficiency of a well-proportioned single-riveted butt joint with double covers is, judging from the few experiments on the subject, somewhat less than that of a double-riveted lap joint—say 55 per cent. of the strength of the solid plate, and that only when the rivets are pitched so that the joint will break indifferently by shearing the rivets or tearing the plate. The efficiency of double-riveted and double-covered butt joints, if well proportioned, may probably be calculated at about 66 per cent. of the strength of the solid plate, unless the plates exceed $\frac{5}{8}$ inch in thickness, when the efficiency of the joint will be somewhat less. The joints in the tension flanges and tension bracing of girders are generally triple or quadruple-riveted, and the transverse pitch of the rivets generally ranges from 3 to 6 inches, so that the efficiency of the joints varies according to the design; but it can scarcely exceed 80 per cent. of the strength of the solid plate, and hence it happens in girder-work that the increased width of flange in consequence of the rivet holes adds seldom less than 20 per cent., and sometimes much more, to the theoretic weight of a tension flange—that is, its weight calculated on the imaginary hypothesis that it is made of solid iron without joints or perforations. As the covers are generally triple or quadruple-riveted, their length is considerable, and their weight forms an important addition, in many cases over 12 per cent., to that of the theoretic solid flange.

10. *Contraction of Rivets and resulting Friction of Plates.*—Rivets contract in cooling and draw the plates together with such force that the friction produced between their surfaces is generally sufficient to prevent them from sliding over each

other so long as the stress lies within limits which are not exceeded in ordinary practice, and in this case the rivets are not subject to shearing stress. From experiments made during the construction of the Britannia Tubular bridge it appears that the amount of this friction is rather variable.* In one experiment, with a $\frac{7}{8}$-inch rivet passing through three plates, and therefore in double shear, it amounted to 5·6 tons, in another, with a $\frac{7}{8}$-inch rivet and two plates lap-jointed, with $\frac{5}{16}$-inch washers next the rivet heads, it reached 4·73 tons, while in a third experiment, with three plates and $\frac{7}{8}$-inch rivet, with $\frac{1}{4}$-inch washers next the rivet heads, making the shank of the rivet $2\frac{7}{8}$-inch long, the middle plate supported 7·94 tons before it slipped. In these experiments the hole in one or both plates was oval, and the slipping took place suddenly. In subsequent experiments made by the Admiralty,† one plate was riveted between two others, and the friction with $\frac{1}{2}$-inch plates and $\frac{3}{4}$-inch rivets amounted to 4·6 tons per rivet, and with $\frac{7}{8}$ plates and 1-inch rivets, it amounted to 5·6 tons per rivet. In these experiments, rivets with pan heads and *conical* points had the advantage over snap points, and countersunk riveting caused much less friction than other systems. This agrees with the author's experience, and Hutchinson also says it is well known among practical boiler makers that iron rivets finished with a cup-shaped snap are not so tight as those hammered until they are nearly cold, and finished off without the use of the snap.‡ Later Admiralty experiments, however, made at Pembroke Dockyard with steel rivets, do not agree with the above, for they showed "that on the whole the friction is greatest for the countersunk rivets."§ In Professor Kennedy's experiments on $\frac{5}{8}$-inch

---

* Clark's Tubular Bridges, p. 393.
† Wilson on Steam Boilers, p. 58.
‡ Hutchinson's Girder-making, p. 62.
§ Wildish, Trans. Inst. Nav. Arch., 1885, p. 190.

steel plates, 11 inches wide, lap-jointed and single-riveted with seven ¾-inch steel rivets and hand-riveted, the mean at which *visible* slipping occurred was about 23·5 per cent. of the breaking load of the joint, but in some cases it was visible very much sooner than this;[*] and other experiments by Prof. Kennedy on double-riveted steel joints, and some made by the Admiralty officials also,[†] indicate that the friction of joints made with machine riveting is much greater than that of joints made with hand riveting. Thus, in Professor Kennedy's experiments the friction of double-riveted steel joints made with hydraulic riveting varied from 34 to 57 per cent. of the breaking load, and was nearly double that of similar joints made with hand riveting.[‡] The hammers used in his experiments, however, were rather light—only 4 lbs. weight, and steel rivets require heavier hammers than iron rivets, in order to knock them down quickly and finish the points before they become black. For instance, 8 lb. hammers are used for hand riveting in the ship yard of Messrs. Harland and Wolff. Notwithstanding the lightness of his hammers, the ultimate strength of the hand-riveted steel joints in Professor Kennedy's experiments was quite as great as that of joints hydraulically riveted. The friction of the plates is an important factor in the staunchness of boilers and, as it is usual to test them hydraulically to double their working pressure, the joints should be designed so that this water test, as well as the expansion and contraction due to changes of temperature, will not cause the joints to slip. Though the friction of riveted plates may be sufficient to convey the normal working load without subjecting the rivets to a shearing stress, it does not follow, nor do experiments indicate, that the *ultimate* strength of a riveted joint is

---

[*] Proc. Inst. M. E., 1881, p. 228.
[†] Wildish, Trans. Inst. Nav. Arch., 1885, p. 190.
[‡] Proc. Inst. M. E., 1885.

increased by this friction. On this subject the following instructive remarks occur in the memorandum issued by the Board of Trade for the use of their surveyors in connection with steel riveting:*—

"It has been usual in some quarters when considering the ultimate stress borne by riveted joints to attribute considerable importance to the friction between the plates caused by the force with which they are held together by the rivets. Whatever value may justly be attached to this at the ordinary working stress, an inspection of the riveted joint when being tested to destruction effectually dispels all idea of the ultimate stress being in the slightest degree affected by it, owing to the extent to which the joint gapes, as shown in the annexed sketch:—

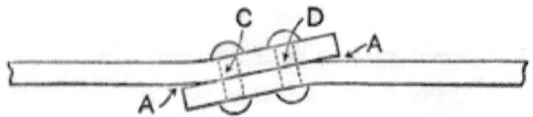

"It is interesting to note that the higher the stress sustained, as in the case of riveted joints with drilled holes, the greater the amount of opening is observed at A. In the case of some 1-inch treble-riveted lap-joints which have been tested for the Board of Trade, the opening has in some cases been as much as $\frac{5}{16}$-inch. It is scarcely necessary to insist upon the fact that under such conditions the ultimate stress of the material either at C or D, where fracture occurs, cannot be increased by any amount of friction which existed at the time f construction. It may not be amiss to remark that even in the case of riveted joints fitted with double butt-straps the bending of the rivets causes the butt-straps to open, and apparently destroys all friction between the plates and butt-straps." When several plates are riveted together with numerous rivets, as in the piled flanges of girders, this slipping

*Merchant Shipping Experiments on Steel, p. 32.

does not occur, for Mr. Baker experimented on two wrought-iron girders, with 5 and 8 plates respectively in their flanges, each of 20 feet span and 2 feet deep, which he tested to failure.* There was no movement in the rivets, and the pile of plates behaved almost like a welded mass of iron, and Mr. Baker states " that he had invariably found that badly punched girders, with the holes partly blind and the rivets tight but not filling the holes, deflected neither more nor less than the most accurately drilled work."

11. *Bearing Area of Iron Rivets.*—Fox's experiments on the eye-bars of suspension bridges indicate that the bar will tear across the eye when the bearing pressure of the pin reaches 40 tons per square inch;† but, as already stated, there seems some evidence to show that when the crushing pressure of a rivet in single shear, that is, its pressure against the plate in a lap joint, exceeds 30 tons per square inch, it reduces the tenacity of the plate; and this, if correct, places a superior limit to the proper diameter of rivets, for when the thickness of the plate remains constant, the shearing area of the rivets increases much faster than their bearing area.‡ Latham says—" If one-inch rivets in half-inch plates be subjected to a calculated average testing stress of 7 tons to the square inch of effective bearing area, some will be found marked by the plates cutting them in a few days," and he adopts 5 tons per square inch of bearing surface as the safe working pressure of rivets against plates.§ It seems probable, however, as Professor Unwin suggests,‖ that some of the rivets in Latham's experiments bore at first harder than the others, and that a minute yielding of one or two rivets merely

---

* Proc. Inst. C. E., Vol. LXI., p. 194.
† Proc. Roy. Soc., Vol. XIV., p. 139.
‡ The bearing area of a rivet is measured by the product of its diameter by the thickness of the plate against which it bears.
§ Latham on Bridges, pp. 18, 24.
‖ Unwin's Iron Bridges, p. 105.

brought the rest to a bearing, thus equalising the stress throughout the joint, for Fairbairn's experiments for the Iron Plate Committee show that iron plates are not indented with a less pressure than 11 tons per square inch,* and in later experiments by Tangye no impression was made on Low Moor iron by a pressure of 20 tons per square inch.† Latham's rule for bearing pressure agrees closely with the usual American practice with pinned girders, as engineers in the United States generally limit the bearing pressure of wrought-iron pins in eye-bars to from 10,000 to 12,000 lbs. (4·46 to 5·36 tons) per square inch on the projected area of the pin.‡ American engineers, however, frequently adopt a much lower shearing stress for rivets than is usual in English practice, for they often limit the shearing stress of rivets to from 6,000 to 6,500 lbs. (2·68 to 2·9 tons) per square inch, with the same bearing pressure as they allow for pins.§ Other American engineers, however, follow English practice closely, and specify that the area of rivets shall not be less than the sectional area of the joined pieces,‖ and Shaler Smith adopts 4·46 tons per square inch for the shearing stress of pins, bolts, and rivets.¶ When a rivet is in double shear, the bearing pressure is frequently much greater than Latham's rule allows, unless, indeed, the friction of the plates prevents this bearing pressure from coming into action. If, for example, a butt-joint be made by riveting $\frac{3}{8}$-inch plates between double covers with rivets exactly $\frac{3}{4}$-inch in diameter, the rivet shearing area on each side of the central plate = 0·44 square inches, and its

---

* Unwin's Iron Bridges, p. 103.

† Clark's Manual of Rules for Engineers, p. 582.

‡ Proc. Inst. C. E., Vol. LIV., p. 197, and Bender on Pins used in Bridges, p. 15.

§ Proc. Inst. C. E., Vol. LXXVII., p. 263; Thurston's Materials of Eng., Part II., p. 641; and Dubois' Strains on Framed Structures, p. 376.

‖ Proc. Inst. C. E., Vol. LIV., p. 197.

¶ Trans. Am. Soc. C. E., 1880, p. 139.

bearing area against the central plate = $\frac{3}{4}'' \times \frac{3}{8}'' = 0.28$ square inches nearly. Consequently Latham's rule would permit a working pressure of only $5 \times 0.28 = 1.4$ tons, whereas the working stress (calculated from the shearing area according to the usual English limit of 4·5 tons shearing stress per square inch in single shear and $1\frac{3}{4}$ times as much in double shear) $= 0.44 \times 1.75 \times 4.5 = 3.46$ tons, or more than twice what Latham's rule allows. The bearing pressure of the links of a common chain against each other often far exceeds the bearing pressure of rivets by any of the foregoing rules.

12. *Strength of Iron Rivets in Tension.*—Rivets are sometimes used for supporting weights, or for holding pieces together by their heads and points, in which case the rivets are subject to tensile in place of shearing stress, and their strength depends on the force required to pull off the rivet heads. In some experiments made by the author with $\frac{3}{4}$-inch iron rivets with pan heads and hand-made snap points in punched holes, the heads or points flew off with an average pull of 7 tons per rivet, or 12·32 tons per square inch of rivet area, and he adopts a factor of safety of 5 for this class of work in girders. Shaler Smith limits the stress on rivets in tension to 2·23 tons per square inch.*

13. *Efficiency of riveted Iron Joints.*—The following is a summary of the conclusions already formed respecting the efficiency of riveted joints with iron plates varying from $\frac{5}{16}$ to $\frac{5}{8}$ inch in thickness. With thicker plates the efficiency will probably be much less. Calling the total tensile strength of the original solid plate 100, the efficiency of various joints are as follows, and that only, provided the pitch of the riveting is so arranged that the joint is on the point of giving way from the tearing of the plate or the shearing of the rivets indifferently. Otherwise from 10 to 20 per cent. may be taken off the percentage of efficiency given in the table. It

* Trans. Am. Soc. C. E., 1880, p. 139.

will also be recollected that the covers of single-covered butt joints should be thicker than the plates they connect (§ 9).

TABLE XVII.—*Relative Efficiency of Iron Joints of various kinds.*

	Efficiency, per cent.
Original solid plate,	100
Lap joint, single-riveted, punched,	45
Do. do. drilled,	50
Do. double-riveted,	60
Butt joint, single cover, single-riveted,	45 to 50
Do. do. double-riveted,	60
Do. double cover, single-riveted,	55
Do. do. double-riveted,	66
Tension flanges of girders, triple or quadruple riveted,	70 to 80

Future experiments may modify the foregoing percentages, which have been deduced from experiments by various persons differing widely in their individual results.

Ex. What is the strength per running inch of a single-riveted lap joint with ¼ inch plates and punched holes if the tensile strength of the iron is 21 tons per square inch? Here the tensile strength of the solid plate is 10·5 tons per running inch, and we have,

Answer. Strength of joint per inch run = ·45 × 10·5 = 4·7 tons, of which, if a marine boiler, one-fifth, = 0·94 tons, will be the safe working stress.

14. *Theoretic Proportions of Joints.*—The theoretic proportions of riveted joints may be found as follows:—

Let d = the diameter of the rivet hole in inches,

t = the thickness of the plate in inches,

p = the transverse pitch in inches,

f = the tensile strength per square inch of the net plate area between the holes,

s = the shearing strength per square inch of the rivet's section in single shear,

c = the crippling pressure per square inch of the rivet against the plate—*i.e.*, the pressure of the rivet that reduces the capacity of the rivet to resist shearing or that of the plate to resist tearing (see §§ 9 and 11).

(a). To find the theoretic diameter of a rivet in single or in double shear. Equate the crippling pressure of the rivet to its shearing strength as follows:—

For *single shear*, $cdt = \cdot 7854 d^2 s$,

whence, $$d = \frac{ct}{\cdot 7854 s} \text{ and } \frac{d}{t} = \frac{c}{\cdot 7854 s} \quad (3)$$

Ex. 1. What is the theoretic diameter of the rivet hole in a ½ inch punched iron plate in a lap joint, either single or double riveted?

Here, $c = 30$ tons per square inch (see § 9),
$t = \frac{1}{2}$ inch,
$s = 19$ tons per square inch for punched holes (see p. 9).

Answer (eq. 3). $d = \frac{ct}{\cdot 7854 s} = \frac{30 \times \cdot 5}{\cdot 7854 \times 19} = 1 \cdot 005$ inches.

For *double shear*, the shearing strength of an iron rivet = 1·75 times that in single shear (p. 10), and we have
$$cdt = 1 \cdot 75 \times \cdot 7854 d^2 s$$
whence, $$d = \frac{ct}{1 \cdot 75 \times \cdot 7854 s} = \frac{ct}{1 \cdot 374 s} \text{ and } \frac{d}{t} = \frac{c}{1 \cdot 374 s} \quad (4)$$

Ex. 2. What is the theoretic diameter of the rivet hole in a ½ inch punched iron plate in a butt joint with double covers?

Here, $c = 40$ tons per square inch (see § 9),
$t = \frac{1}{2}$ inch,
$s = 19$ tons per square inch for punched holes.

Answer (eq. 4). $d = \frac{ct}{1 \cdot 374 s} = \frac{40 \times \cdot 5}{1 \cdot 374 \times 19} = \cdot 77$ inches.

(b). To find the theoretic pitch in single-riveted joints, either in single or double shear. Equate the crippling pressure of one rivet to the tearing strength of the net plate area between two holes as follows:—

$$cdt = ft(p - d)$$

whence, $$p = \frac{c + f}{f} d \text{ and } \frac{p}{d} = \frac{c + f}{f} \quad (5)$$

Aliter.

Equate the shearing strength of one rivet to the tearing strength of the net plate area between two holes, as follows:

OF RIVETED JOINTS.

For *single shear*, $\cdot 7854 d^2 s = ft(p-d)$

whence,
$$p = \frac{\cdot 7854 d^2 s + ftd}{ft} \qquad (6)$$

For *double shear*, $1\cdot 75 \times \cdot 7854 d^2 s = ft(p-d)$

whence,
$$p = \frac{1\cdot 374 d^2 s + ftd}{ft} \qquad (7)$$

Ex. 3. What is the theoretic pitch for a ½ inch punched iron plate in a single-riveted lap joint, the tensile strength of the solid plate being 22 tons per square inch, but reduced 20 per cent. by punching and the bending of the joint ? (see § 7.)

Here, $c = 30$ tons per square inch (see § 9),
$t = \frac{1}{2}$ inch,
$d = 1$ inch (see ex. 1),
$s = 19$ tons per square inch for punched holes,
$f = 0\cdot 8 \times 22$ tons $= 17\cdot 6$ tons per square inch (see § 7).

Answer (eq. 5). $p = \dfrac{c+f}{f} d = \dfrac{30 + 17\cdot 6}{17\cdot 6} = 2\cdot 7$ inches.

Aliter (eq. 6). Answer. $p = \dfrac{\cdot 7854 d^2 s + ftd}{ft} = \dfrac{(\cdot 7854 \times 19) + (17\cdot 6 \times \cdot 5)}{17\cdot 6 \times \cdot 5} = 2\cdot 7$ in.

Ex. 4. What is the theoretic pitch for a ½ inch punched iron plate in a single-riveted butt joint with double covers, the tensile strength of the solid plate being 22 tons per square inch, but reduced by punching to 20 tons per square inch in the perforated plate ?

Here, $c = 40$ tons per square inch (see § 9),
$t = \frac{1}{2}$ inch,
$d = \cdot 77$ inches (see ex. 2),
$s = 19$ tons per square inch for punched holes,
$f = 20$ tons per square inch.

Answer (eq. 5). $p = \dfrac{c+f}{f} d = \dfrac{40 + 20}{20} \times \cdot 77 = 2\cdot 3$ inches.

Aliter (eq. 7). Answer. $p = \dfrac{1\cdot 374 d^2 s + ftd}{ft} =$

$\dfrac{(1\cdot 374 \times \overline{\cdot 77}|^2 \times 19) + (20 \times \cdot 5 \times \cdot 77)}{20 \times \cdot 5} = 2\cdot 31$ inches.

(*c*). To find the theoretic pitch in double-riveted joints, either in single or double shear. Equate the crippling pressure of two rivets to the tearing strength of the net plate area between two holes as follows:—

$$2cdt = ft(p-d)$$

whence,
$$p = \frac{2c+f}{f} d \text{ and } \frac{p}{d} = \frac{2c+f}{f} \qquad (8)$$

Aliter.

Equate the shearing strength of two rivets to the tearing strength of the net plate area between two holes, as follows:

For *single shear*, $2 \times \cdot 7854 d^2 s = ft (p - d)$

whence,
$$p = \frac{1 \cdot 57 d^2 s + ftd}{ft} \qquad (9)$$

For *double shear*, $2 \times 1 \cdot 75 \times \cdot 7854 d^2 s = ft (p - d)$

whence,
$$p = \frac{2 \cdot 748 d^2 s + ftd}{ft} \qquad (10)$$

Ex. 5. What is the theoretic pitch for a ½ inch punched iron plate in a double-riveted lap joint, the tensile strength of the solid plate being 22 tons per square inch, but reduced by punching to 20 tons in the perforated plate?

Here, $c = 30$ tons per square inch (see § 9),
$f = 20$ tons per square inch,
$t = \frac{1}{2}$ inch,
$s = 19$ tons per square inch,
$d = 1$ inch (see ex. 1).

Answer (eq. 8). $p = \dfrac{2c + f}{f} d = \dfrac{2 \times 30 + 20}{20} = 4 \cdot 0$ inches.

Aliter (eq. 9). Answer. $p = \dfrac{1 \cdot 57 d^2 s + ftd}{ft} =$

$$\frac{(1 \cdot 57 \times 19) + (20 \times \cdot 5)}{20 \times \cdot 5} = 3 \cdot 98 \text{ inches.}$$

Ex. 6. What is the theoretic pitch for a ½ inch punched iron plate in a double-riveted butt joint with double covers, the tensile strength of the solid plate being 22 tons per square inch, but reduced by punching to 20 tons in the perforated plate?

Here, $c = 40$ tons per square inch (see § 9),
$t = \frac{1}{2}$ inch,
$f = 20$ tons per square inch,
$d = \cdot 77$ inches (see ex. 2),
$s = 19$ tons per square inch.

Answer (eq. 8). $p = \dfrac{2c + f}{f} d = \dfrac{2 \times 40 + 20}{20} \times \cdot 77 = 3 \cdot 85$ in.

Aliter (eq. 10). Answer. $p = \dfrac{2 \cdot 748 d^2 s + ftd}{ft} =$

$$\frac{(2 \cdot 748 \times \cdot 77{\mid}^2 \times 19) + (20 \times \cdot 5 \times \cdot 77)}{20 \times \cdot 5} = 3 \cdot 87 \text{ inches.}$$

PART II.

STEEL PLATES AND STEEL RIVETS.

For most of the following information on steel riveting the author is indebted to three sources:—Firstly—" Merchant Shipping experiments on steel," a memorandum issued by the Board of Trade in 1881 for the use of their Surveyors, and prepared by Mr. Traill, Engineer Surveyor in chief, and his assistants, Messrs. Richards and Sampson. Secondly—Prof. Kennedy's " Results of experiments on riveted joints," made for the Research Committee of the Institution of Mechanical Engineers, and published in their Proceedings for 1881, 1882, and 1885. Thirdly—Mr. Moberly's papers " On tests of riveted joints," published in 1882 and 1883 in the Proceedings of the Institution of Civil Engineers, Vols. LXIX. and LXXII.

15. *Shearing Strength of Bar Steel by direct experiments.*— Few direct experiments have hitherto been made on the shearing strength of steel bars, and Table XVIII. contains the principal ones yet published, and it will be observed that the average shearing strength of the steel bars in this table is approximately 84 per cent. of their average tensile strength, but this ratio varies widely in individual experiments.

Mr. Baker says that " the steel rivets in the Forth bridge have a tensile strength of about 27 tons, an elongation of about 30 per cent., and a shearing resistance of from 22 to 24 tons per square inch."* Extra mild rivet steel seriously reduces the efficiency of a joint, and a shearing strength of not less than 24 tons per square inch is probably the correct standard to seek for.

* Engineering, August, 1884, p. 214.

TABLE XVIII.—*Tensile and Shearing strengths of Steel Bars, derived from direct experiments.*

Experimenter.	Maker of Steel and Remarks.	Diameter of bars.	Tensile strength per square inch.	Elastic Limit per square inch.	Ultimate Set after fracture in 10 inches.	Shearing strength per square inch.	Reference.
		inches.	tons.	tons.	Per cent.	tons.	
Greig & Eyth	John Brown & Co.'s mild steel, in double shear.	$\frac{7}{8}$	28·83	—	20·6	22·18	Proc. I. M. E., 1879, p. 292.
Kennedy	Landore Siemens Steel Co.'s mild rivet steel, in single shear.	$1\frac{1}{8}$	29·17	19·8	20·9	23·41	*Idem*, 1881, pp. 239, 242.
Do.	Do. do.	$1\frac{3}{8}$	30·73	20·65	19·9	26·63	Do.
Do.	Do. do.	$1\frac{1}{16}$	27·46	21·09	23·3	24·35	Do.
Do.	Do. do.	$\frac{3}{4}$	29·03	18·62	—	26·3	*Idem*, p. 256.
	Mean	-	29·04	—	—	24·6	

16. *Shearing Strength of Steel Rivets in the Joint.*—Table XIX. contains the results of experiments made by Kirkaldy for Mr. Moberly on single-riveted lap joints with *punched* holes.*

TABLE XIX.—*Shearing strength of Steel Rivets in single shear, derived from experiments on single-riveted Lap Joints with* PUNCHED *holes, broken by shearing* (Moberly).

Mode of riveting.	Tensile strength per square inch of rivet.	Shearing strength per square inch of rivet area.	Number and size of rivets.		Remarks.
			No.	Diam.	
	tons.	tons.		in.	
Machine	28·3	23·2	5	1	Plates $\frac{7}{16}$-inch Landore S steel, and rivets Landore rivet steel. Plates were not annealed after punching.
Do.	28·3	23·2	5	1	
Mean	-	23·2			

* Proc. Inst. C. E., Vol. LXIX., pp. 359, 361.

Table XX. contains the results of experiments by Professor Kennedy on single-riveted lap joints with *drilled* holes.*

TABLE XX.—*Shearing strength of Steel Rivets in single shear, derived from experiments on single-riveted Lap Joints with* DRILLED *Holes, broken by shearing* (Kennedy).

Mode of riveting.	Thickness of plate.	Number and size of rivets.		Original tensile strength per square inch.		Shearing strength per square inch of rivet area.	Bearing pressure per square inch.	Remarks.
		No.	Diam.	Plate.	Rivet.			
	in.		in.	tons.	tons.	tons.	tons.	
Hand	3/4	2	1·1	27·27 with grain, and 26·93 across grain	—	22·01	26·29	
,,	,,	,,	,,		—	21·58	25·55	
,,	,,	,,	1		—	21·84	23·59	
,,	3/8	,,	,,	29·97	29·12	16·63	35·01	
,,	,,	,,	,,	,,	,,	17·29	36·18	
,,	,,	,,	3/4	,,	,,	24·31	39·64	Plates Landore SS steel. Rivets Landore mild rivet steel. The shearing strength of the rivet steel by direct experiment varied from 23·4 to 26·63 tons per square inch (see table XVIII.).
,,	,,	3	,,	29·91	29·03	22·01	35·83	
,,	,,	,,	,,	,,	,,	21·37	35·18	
,,	,,	7	,,	29·33	,,	21·46	33·17	
,,	,,	,,	,,	,,	,,	18·23	31·00	
,,	,,	,,	,,	,,	,,	22·84	33·46	
,,	,,	,,	,,	,,	,,	22·08	35·56	
,,	,,	,,	,,	,,	,,	20·62	31·70	
,,	,,	,,	,,	,,	,,	22·09	39·71	
				Mean	-	21·03, or 22·02 omitting experiments 4, 5 and 10, in which the steel seems exceptionally soft.		

This table seems to indicate that the shearing strength of large-sized steel rivets (1 inch and upwards) is somewhat less than that of smaller sizes. In experiments made at Pembroke

* Proc. Inst. M. E., 1881, pp. 248, 251, 254, 256, 717.

Dockyard, the shearing strengths of ¾, ⅞, and 1-inch rivets were substantially in proportion to their respective areas.* Comparing Tables XIX. and XX., we find that the shearing strength of steel rivets in drilled holes is less than in punched holes, in this respect agreeing with iron rivets. It would be desirable, however, to repeat the experiments with the same rivet steel in both punched and drilled holes, as the shearing strengths of different specimens of rivet steel vary greatly. Professor Kennedy found that the size of the rivet heads and points played a most important part in the strength of single-riveted lap joints, an increase of about one-third in the weight of the rivets, going to the heads and points, increasing their shearing strength in some of his experiments from a little over 20 tons to over 22 tons per square inch. This additional strength he attributes to the prevention of so great tensile stress in the rivets through distortion of the plates.† Mr. Webb also likes to have deep heads to the rivets in locomotive steam boilers, so that they should not give way, or curl up round the edge. Mr. Moberly tried some experiments on single-riveted lap joints made with punched curved plates, similar to the ring seams of a boiler, and he found that the curvature greatly prevented the bending of the joint and increased the shearing strength of the rivets from 23·28 to 24 tons, that is about 3 per cent.‡

Table XXI. contains the results of experiments made for Lloyd's Committee and described by Mr. Martell, and others made by Kirkaldy for Mr. Moberly, on double-riveted lap joints with *punched* holes, and it will be observed that both the tensile and shearing strengths of the rivets in Mr. Moberly's experiments were above their average strengths in the other tables.

* Wildish, Trans. Inst. Nav. Arch., 1885, pp. 187, 189.
† Proc. Inst. M. E., 1885.
‡ Proc. Inst. C. E., Vol. LXIX., p. 353.

TABLE XXI.—*Shearing strength of Steel Rivets in single shear, derived from experiments on double-riveted Lap Joints with* PUNCHED *Holes, broken by rivets shearing.*

Mode of riveting.	Original tensile strength per sq. inch.		Maker and size of Plate.	No. and size of rivets sheared, or size of hole.		Shearing strength per square inch of Rivet Area.	Experimenter.	Authority and Reference.
	Plate.	Rivet.		No	Diam.			
	tons.	tons.	inch.		inch.	tons.		
—	—	—	$12\frac{3}{8} \times 3\frac{3}{8}$	7	$\frac{13}{16}$	24·1	—	Martell, Trans. Inst. Nav. Arch., 1878, p. 14.
Machine	—	29·7	Landore—17·15 × 0·44	6	0·81	25·8	Kirkaldy	Moberly, Proc. Inst. C. E., Vol. LXXII, pp. 246, 247.
Do.	—	29·7	,, 17·15 × 0·44	6	0·81	25·8	Do.	Do.
Do.	—	32·6	,, 15·25 × 0·32	6	0·68	26·9	Do.	Do.
Do.	—	32·6	,, 15·25 × 0·32	6	0·68	25·6	Do.	Do.
					Mean,	25·6		

Table XXII. contains the results of various experiments on double and triple-riveted lap joints with *drilled* holes, and it corroborates the inference that the shearing strength of large-sized steel rivets is slightly less than that of smaller sizes. It also proves that the shearing strength of rivet steel varies greatly.

Table XXII.—*Shearing strength of Steel Rivets in single shear, derived from experiments on double or triple-riveted Lap Joints with* DRILLED *Holes, broken by rivets shearing.*

Mode of riveting.	Original tensile strength per sq. inch.		Size of Plate.	No. and size of rivets sheared, or size of hole.		Shearing strength per sq. in. of Rivet Area.	Bearing pressure per square inch.	Remarks.	Experimenter.	Authority and Reference.
	Plate.	Rivet.		No.	Diam.					
	tons.	tons.	inch.		inch.	tons.	tons.			
—	—	—	$12\frac{1}{8} \times \frac{13}{16}$	7	$\frac{13}{16}$	24·25	—	Double-riveted.	—	Martell, Trans. Inst. Nav. Arch., 1878.
—	—	—	$10 \times \frac{9}{16}$	9	$\frac{3}{4}$	27·4	—	Triple-riveted.	—	Do.
—	—	—	$10 \times \frac{9}{16}$	9	$\frac{3}{4}$	26·7	—	Do.	—	Do.
—	—	—	$11\frac{1}{4} \times \frac{3}{4}$	6	$1\frac{1}{16}$	19·7	—	Do.	—	Do.
—	—	—	$11\frac{1}{4} \times \frac{3}{4}$	6	$1\frac{1}{16}$	19·7	—	Do.	—	Do.
—	—	—	$13\frac{1}{8} \times \frac{3}{4}$	9	1	22·2	—	Do.	—	Do.
—	—	—	$13\frac{1}{8} \times \frac{3}{4}$	9	$1\frac{1}{8}$	21·9	—	Do.	—	Do.
—	—	—	$11\frac{1}{4} \times \frac{3}{4}$	6	$1\frac{1}{32}$	22·0	—	Do.	—	Do.
—	—	—	$11\frac{1}{4} \times \frac{3}{4}$	6	$1\frac{1}{32}$	23·3	—	Do.	—	Do.
Machine.	31·7	31·2	$14 \times \frac{1}{2}$	21	0·45	25·1	—	Triple-riveted, Steel Co. of Scotland.	Kirkaldy	Merchant Shipping Experiments on Steel, p. 83.
Do.	29·1	30·6	$13\frac{1}{2} \times \frac{1}{2}$	18	0·64	25·5	—	Do.	Do.	Do.
Do.	30·4	28·6	$14 \times \frac{3}{4}$	12	0·95	23·9	—	Do.	Do.	Do., p. 85
Do.	27·5	30·2	11×1	9	1·08	24·1	—	Do.	Do.	Do.
Hand.	29·97	—	$10 \times \frac{3}{8}$	7	0·8	24·8	41·4	Double-riveted, Landore steel.	Kennedy	Kennedy, Proc. I.M.E., 1885.
Hydraulic.	29·97	—	$10 \times \frac{3}{8}$	7	0·8	21·57	35·98	Do.	Do.	Do.
Do.	28·46	—	$11 \times \frac{3}{4}$	7	1·1	21·57	25·44	Do.	Do.	Do.
Do.	28·46	—	$11 \times \frac{3}{4}$	7	1·1	20·85	24·53	Do.	Do.	Do.
				Mean -		23·2				

Comparing Tables XXI. and XXII., it would again appear as if the shearing strength of steel rivets in punched holes is greater than in drilled holes. Comparing also the experiments on single-riveted lap joints (Tables XIX. and XX.) with those on double or triple-riveted lap joints (Tables XXI. and XXII.), we find that the severe bending of the joint in single-riveted laps seriously reduces the shearing strength of the rivet.

TABLE XXIII.—*Shearing strength of Steel Rivets in single shear, derived from experiments on Butt Joints with single covers of same thickness as Plates and double-riveted. Holes in Plates, some* PUNCHED, *and some* PUNCHED *and* COUNTERSUNK. *Holes in Covers* DRILLED (Wildish).

	Mean shearing strength per square inch of rivet area.			
	Pan head and countersunk point. Countersink 1/16 inch less than full thickness of Plate.	Pan head and countersunk point. Countersink extended right through Plate.	Countersunk head and point.	Pan head and snap point.
	tons.	tons.	tons.	tons.
¾-in. rivets in ½-in. plates,	21·56	23·17	No result obtained. Some rivets drawn through, others sheared.	21·07
⅞-in. rivets in ⅝-in. plates,	22·52	22·13		21·55
1-in. rivets in ¾-in. plates,	23·72	22·55	23·59	21·58
Means, -	22·6	22·62	—	21·4

Table XXIII. gives the mean results of several experiments made at the Admiralty Dockyard, Pembroke, on the shearing strength of steel rivets in double-riveted butt joints with single covers of the *same* thickness as the plates, such as are used at the butt joints in the outer plating of ships. The plates were punched and the straps drilled, as is the practice in the Naval yards. Some of the plates were countersunk for countersunk rivets, others had snap-pointed rivets; and Mr. Wildish states that the tensile strength of the bars from which the rivets were made may be taken at 28 tons per square inch.* It will be observed that the shearing strength of rivets with snap points was less than that of rivets with countersunk points; why, is not very apparent.

The foregoing table corroborates the opinion of Mr. Denny, whose experience as a shipbuilder is very great, and who gives the results of his experiments in the following words: "Taking a fair view of the matter, I do not think it would be prudent to assume in ship-riveting—which must in the most important part, the skin, be done by hand—a higher shearing stress (for steel) than 22 tons per square inch of area, against, say 19 tons for an iron rivet."†

The shearing strength of steel rivets in double shear, in three experiments (see Table XXIV.) made by Kirkaldy for Mr. Moberly, on double-riveted and double-covered butt joints, with $\frac{9}{16}$-inch plates, differed according as the holes were punched, or punched and bored afterwards; and it will be observed that these experiments are at variance with the previous inference that rivets in punched holes are stronger than those in drilled holes, and further experiments are wanted to decide this question. In other respects, however, they are somewhat anomalous, for there does not appear sufficient reason why the shearing unit-strength should be so much greater in punched and drilled holes than in punched and bored holes.‡

Messrs. Greig & Eyth made fifteen experiments§ on the shearing strength of $\frac{7}{8}$-inch rivets in double shear in single-riveted butt joints with double covers and *drilled* holes. The rivets were both hand and machine-riveted, and their shearing

---

* Wildish. Trans. Inst. Nav. Arch., 1885, p. 189.
† Trans. Inst. Nav. Arch., 1880, p. 192.
‡ Proc. Inst. C. E., Vol. LXIX., p. 358.
§ Proc. Inst. M. E., 1879, p. 294.

strength varied from 22·5 to 26·3 tons per square inch, the mean being **24·1** tons.

TABLE XXIV.—*Shearing strength of Steel Rivets in double shear in double-riveted and double-covered butt joints with holes formed in various ways* (Moberly).

Shearing Strength per square inch of Rivet Area, $\frac{9}{16}$ in. plates.			
Punched Holes, $\frac{13}{16}$.	Punched, $\frac{3}{4}$, and bored, $\frac{13}{16}$.	Punched, $\frac{3}{4}$, and drilled, $\frac{3}{4}$.	Mean.
tons.	tons.	tons.	tons.
23·8	24·2	25·8	**24·6**

Mr. Denny made several experiments on the shearing strength of single steel rivets in double shear with *drilled* holes, the results of which are given in the following table:*—

TABLE XXV.—*Shearing strength of Steel Rivets in double shear with* DRILLED *Holes* (Denny).

Size of Rivet and kind of Steel.	Machine or Hand-riveted.	Shearing strength per square inch.	Remarks.
$\frac{3}{4}$-inch rivets made from homogeneous steel.	Machine	tons. 24·6	Mean of 4 experiments, varying from 24 to 25·1 tons.
	Hand	23·3	Mean of 4 experiments, varying from 20·5 to 25·6 tons.
$\frac{3}{4}$-inch rivets made from scrap steel.	Machine	23·3	Mean of 4 experiments, varying from 22·8 to 24·1 tons.
	Hand	21·6	Mean of 4 experiments, varying from 20·8 to 22·1 tons.

†Trans. Inst. Nav. Arch., 1880, p. 204.

Professor Kennedy concludes that the shearing unit-strength of steel rivets in single and in double shear is the same, provided the lap joints are double-riveted. In single-riveted lap joints, as we have already seen, the shearing strength is greatly reduced by the severe bending of the joint. The previous tables show that the shearing and tensile strengths of rivet steel vary very greatly, and although the shearing strength of rivets is generally about 80 per cent of their tensile strength in drilled joints, and 85 per cent. in punched joints, these ratios cannot be depended on, for Professor Kennedy says that some 35-ton Bessemer steel which he tested by direct experiment had only a shearing resistance of 22 to 23 tons per square inch, and in Mr. Moberly's experiments the tensile strength of ¾-inch steel rivets taken from the same lot varied from 31·5 to 38 tons per square inch.* It would, therefore, be very desirable that steel manufacturers should endeavour to make tough rivet steel with a guaranteed shearing strength of not less than 24 tons per square inch. The Board of Trade rules for marine boilers adopt 23 tons per square inch as the shearing strength of steel rivets in double-riveted lap joints, provided the tensile strength of the rivets is not less than 28 and not more than 32 tons per square inch; and, judging from the various experiments above recorded, it will probably be safe to adopt the following provisional standards for calculating the shearing strength of steel rivets, provided the heads and points of rivets in single-riveted lap joints are of good size. Future experiments may modify this table and show that rivets in punched holes are no stronger than those in drilled holes, especially if the latter have their arrises scraped or very slightly countersunk to remove the sharp film produced by the drill, as is now the general practice in boiler work. :—

* Proc. Inst. C. E., Vol. LXIX., p. 363.

	Standard Shearing strength in tons per square inch of Rivet Section.
Single-riveted lap joints, punched holes,	- 23
Do. do. drilled holes,	- 22
Double-riveted lap joints, punched holes,	- 24
Do. do. drilled holes, -	- 23
Butt joints with single covers of same thickness as plates, double-riveted, snap points,	- 21
Do. do. double-riveted, countersunk points, - -	- 22
Butt joints with double covers, punched holes,	- 24
Do. do. drilled holes,	- 23

17. *Tensile Strength of Perforated Steel Plates (not riveted).*—Experiments show that punching materially injures the tenacity of steel plates when their thickness exceeds $\frac{1}{2}$-inch, the injury being greater in thick than in thin plates, and with hard than with ductile steel; hence the holes in steel plates over $\frac{1}{2}$-inch thick, unless they are countersunk, as in shipbuilding, are generally formed in one of the following ways, which are arranged in their probable order of merit:—

   (1) Drilling.
   (2) Punching and annealing. *
   (3) Punching small and rimering to size.

Mr. White states, that in the Admiralty practice the holes in important parts of the structure were countersunk through the full thickness of the plates, and the material damaged in punching was thus entirely removed. †

The following abstracts of experiments made by Kirkaldy for the Board of Trade on steel plates of different thickness,

---

* Annealing plates is sometimes done by raising them to a dull red heat and covering them with ashes or sand until cold, or better still, leaving the plates in the annealing oven until it cools; covering up, however, is generally omitted, and leaving plates in the oven scarcely ever practised.

† Proc. Inst. C. E., Vol. LXIX., p. 38.

all manufactured by the Steel Company of Scotland, illustrate the effects of different modes of perforation, the holes being ·79 and 1·08 inch in diameter, and it will be seen that, though punching seriously reduced the unit-strength of plates ½-inch thick and upwards, drilling materially increased the unit-strength of all thicknesses (but in a less degree for thick than for thin plates), as did also punching combined with annealing.* These experiments are very valuable, as the specimens were of large size, the plates being 12 inches wide and in most cases perforated with 6 transverse holes at 2 inches pitch when the holes were ·79 inch in diameter, and perforated with 4 transverse holes at 3 inches pitch when the holes were 1 inch in diameter: †—

TABLE XXVI.—*Experiments on the Tensile strength of Steel Plates perforated in different ways, but not riveted* (Board of Trade).

Specimens.	Tensile strength per square inch of net section between holes.			
	¼-in. thick.	½-in. thick.	¾-in. thick.	1-in. thick.
	tons.	tons.	tons.	tons.
Unperforated - - -	31·65	29·15	29·7	27·7
Punched - - -	31·94	27·53	24·6	21·02
Punched and annealed -	33·41	30·75	30·05	27·82
Drilled - - - -	36·21	32·44	31·64	29·12

The foregoing stresses may be usefully expressed in percentages of strength of the solid plate as follows:—

* The process of annealing adopted in the case of the plates so treated was to heat them to a dull red heat and cover them with ashes until cold.—*Merch. Ship. Expts.* p. 12.

† Merchant Shipping Experiments on Steel, p. 13.

TABLE XXVII.—*Relative percentages of strength of Steel Plates perforated in different ways* (Board of Trade).

Specimens.	Unit-strength of net section between holes compared with that of the solid plate (= 100).			
	¼-inch.	½-inch.	¾-inch.	1-inch.
	per cent.	per cent.	per cent.	per cent.
Punched	101·0	94·2	82·5	75·8
Punched and annealed	105·6	105·6	101·0	100·3
Drilled	113·8	111·1	106·4	106·1

The mode in which the ductility of the steel is affected by the treatment it received is shown by the elongation of the holes at the ultimate stress as follows:[*]—

TABLE XXVIII.—*Experiments on the elongation of Holes formed in different ways* (Board of Trade).

Specimens.	Elongation of holes at ultimate stress.			
	¼-inch.	½-inch.	¾-inch.	1-inch.
	per cent.	per cent.	per cent.	per cent.
Punched	11·7	18·5	11·1	4·3
Punched and annealed	27·1	35·1	33·0	29·8
Drilled	24·3	37·0	37·6	33·5

The results above stated were fully corroborated by a second series of experiments on narrower plates (with this exception, that there was not so great a difference between the tensile strength of the drilled specimens and those punched and afterwards annealed), and it was found that "punching small and boring to size," as also "punching and countersinking," increased the unit-strength of the plate much in the same way as drilling did, but the operations of

[*] Merch. Ship. Expts., p. 16.

boring to size or countersinking did not remove the whole effects of punching, for they did not leave the plate the same amount of ductility that drilling did. "The effect of perforating a plate in one or more lines across the direction of stress is to cause lines of weakness at those parts to which also the extension is almost wholly confined. Consequently, owing to the form of the specimen where weakened, and the shortness of the length over which extension occurs as compared with 10 inches (the length of the test samples), the percentage of extension upon that length is very much less than in the unperforated specimen," as shown by the following table, derived from this second series of experiments,[*] which equally well with Table XXVIII. illustrates the way in which the ductility of the steel is affected by the mode of perforation.

TABLE XXIX.—*Experiments on the ultimate extension of solid and perforated Steel Plates* (Board of Trade).

Specimens.	Mean ultimate extension set in 10 inches.			
	¼-inch.	½-inch.	¾-inch.	1-inch.
	per cent.	per cent.	per cent.	per cent.
Unperforated - - -	22·9	30·3	30·6	29·0
Punched - - -	1·7	2·9	2·4	0·75
Punched and annealed	5·3	8·0	8·5	8·0
Punched and bored -	4·2	6·6	6·7	7·5
Drilled - - -	5·8	7·8	9·2	12·0

The small extension of the holes in the specimens, which were bored after being punched, compared with the annealed and drilled specimens, appears to prove that the annulus of metal bored out after punching was scarcely sufficiently large

[*] Merchant Shipping **Experiments on Steel**, p. 23.

to remove the hardening effect of punching. It was also found that punched and countersunk plates, the countersinking extending right through the plate, were as strong as drilled and countersunk plates, but they were not quite as ductile, proving that the hardening effect of the punching had not been totally removed by countersinking.* The foregoing experiments seem to show that in the event of the stress in a boiler constructed of very ductile steel approaching a dangerous limit, warning would probably be given by leakage at the rivet holes if the holes were drilled or the plates properly annealed after punching. †

Professor Kennedy's experiments for the Research Committee of the Institution of Mechanical Engineers, and Admiralty experiments recorded by Mr. Wildish,‡ so far as they go, generally corroborate those made for the Board of Trade, and there can be no doubt therefore that punching does, as a rule, materially reduce the unit-strength of steel plates over ½ inch in thickness, owing, no doubt, to its hardening effect on the annulus of metal round each hole, while other modes of perforation actually increase the tenacity of mild steel. At first sight it seems paradoxical that drilling holes in a ductile plate increases the unit-strength of the metal between the holes, and Professor Kennedy suggests § that it may be due to the contraction of area at the hole altering the usual somewhat irregular flow of the metal under severe stress and rendering it more uniform than in an unperforated plate, so that the stress is more uniformly distributed throughout the whole of

---

* Merchant Shipping Experiments on Steel, 26.

† *Idem*, p. 16.

‡ In the Admiralty experiments punching ½-inch plates with ⅞-inch holes reduced their tensile strength from 28¼ to 22 tons per square inch, and when the holes were drilled the corresponding result was 29½ tons nearly, and in those with the holes punched small and countersunk to the full size and usual taper, the tensile strength was just under 29 tons. Trans. Inst. Nav. Arch., 1885, p. 182.

§ Proc. Inst. M. E., 1881, p. 217.

the fractured area of the perforated plate than throughout the solid plate. Kirkaldy, however, attributes the excess unit-strength due to drilling to the resistance offered to stretching, and consequently to contraction of area, when the breadth of the specimen is reduced by holes, and remarks:—" The ultimate stress borne by a specimen is greatly affected by the hardness or the softness of the material and by the shape of the specimen. The softer the material, the more rapidly does its sectional area become reduced by the specimen stretching, and consequently in the amount of stress sustained. When the breadth of a specimen is reduced to a minimum at one point, a greater resistance is offered to its stretching than when formed parallel for some distance; and as the stretching is checked, so will also the contraction of area, and with it will be an increase in the ultimate stress."[*] Though the apparent tenacity per square inch of net area of a drilled joint with mild steel may thus be greater than that of the solid plate, it is not always safe to reckon on this in calculating the strength of a joint, for steel plates from different manufacturers vary much in ductility and tensile strength.[†] Moreover, the excess tenacity of the plate will probably nearly altogether disappear when the pitch of the rivets is as wide as is usual in girderwork. With reference to the effect of different modes of perforation, Mr. Martell sums up his principal conclusions as follows:[‡]—

Firstly.—" That in plates above $\frac{1}{2}$ an inch in thickness, the loss of strength of iron plates by punching ranged from 20 per cent. to 23 per cent., while in steel plates of the same thickness it ranged from 22 per cent. to 33 per

---

[*] Merch. Ship. Expts., p. 14.
[†] The Board of Trade's instructions to their surveyors limit the makers of steel for marine boilers to 9 firms by name, and state that when the steel is not made by any of these makers the case will receive the special consideration of the Board, and this should be noted by the surveyors.
[‡] Trans. Inst. Nav. Arch., 1878, p. 5.

cent. of the original strength of the plates between the rivet holes. An occasional plate both of iron and steel showed a smaller loss than the minimum here indicated, but they were exceptional cases, so far as these experiments go."

Secondly.—"That by annealing after punching the whole of the lost strength was restored, and in some instances greater relative strength was obtained than existed in the original plates."

Thirdly.—"That the steel was injured only a small distance round the punched holes, and that by riming with a larger drill than the punch, from $\frac{1}{16}$th to $\frac{1}{8}$th of an inch around the holes, the injured part was removed, and no loss of strength was then observable, any more than if the hole had been drilled."

Mr. Martell further observes that the holes in the outside strakes of ships' plating have nearly all the distressed or injured parts around the hole removed by counter-sinking. It will be observed in the experiments recorded in Table XXVI. that punching reduced the unit-strength of all thicknesses of plates, except the $\frac{1}{4}$-inch plates. In the following experiments by Professor Kennedy, *punching* 1-inch holes with 2-inch pitch materially increased the unit-strength of soft $\frac{3}{8}$-inch Landore steel plates, as shown by the following table;* and it was not found to make any difference whether the die was $\frac{1}{32}$ or $\frac{3}{32}$-inch larger in diameter than the punch : †—

---

* Proc. Inst. M. E., 1881, pp. 243 to 246.

† In punching the holes for the Board of Trade's experiments, the dies were larger than the punches by about one-fifth of the diameter, this proportion being adopted as affording a moderate degree of clearance to the punch.—Merch. Ship. expts. on steel, p. 18.

TABLE XXX.—*Experiments on the Tensile strength of Landore* **Steel** *Plates perforated in different ways* (Kennedy).

Specimen.	Tensile strength per square inch of net section.			
	½-inch plates.		⅜-inch plates.	
	1st series.	2nd series.	1st series.	2nd series.
	tons.	tons.	tons.	tons.
Unperforated - -	34·41	29·00	31·45	28·87
Punched - - -	34·74	30·87	34·14	30·53
Drilled - - -	38·17	32·26	35·15	32·00

In these experiments it will be observed that, 1st, the plates did not exceed ⅜-inch in thickness. 2ndly, they were Landore SS steel, a special quality of very mild steel, in which the percentage of carbon is said to be about 0·18; and the late Sir **William** Siemens, who controlled the Landore **Works**, always maintained that punching did not injure mild steel, whereas nearly all experimenters with other brands of steel maintained the reverse to be the case. Hence we may, probably, infer that very mild steel is in this respect somewhat exceptional, for it will be observed in Table XXX. that with somewhat harder steel (34·4 tons in ¼-inch plate) the excess tenacity fell to 0·33 tons per square inch when the holes were punched, but remained as before for drilled holes. 3rdly, the holes were large and close together—1 inch in diameter, with only 1 inch of metal between hole and hole—and the tensile unit-strength of punched steel varies very considerably with the distance apart of the holes. Mons. A. Considére, indeed, concludes from his experiments that the resistance of punched steel may even surpass the normal strength of the unpunched metal in the case of the holes being very near together.[*] He explains this fact by the effect of punching, which, though it hardens and diminishes the ductility of an annulus of metal around each hole, yet increases its tensile strength,

[*] Trans. Inst. Nav. Arch., 1884, p. 296.

much the same way as cold drawing strengthens wire. If now the holes are spaced close together, the annuli of hardened metal round two adjacent holes will touch, or nearly touch each other, and the mean unit-strength of the metal between the holes at the crisis of rupture will be nearly equal to that of the annuli, that is, higher than the normal unit-strength of the plate. If, however, the holes are far apart, the annuli of hardened metal do not stretch so much as the intermediate portions, and the annuli will therefore tear before this intermediate portion has come to its full capacity of tension, thus reducing the mean unit-strength of the metal between the holes at the moment of rupture below that of an unperforated plate. In fact, the metal breaks in detail—first the annuli of hard metal, and then the more ductile intermediate portions.

From the facts recorded in this section we may infer that punching seriously reduces the unit-strength of steel plates exceeding $\frac{1}{2}$-inch in thickness and to a less extent that of $\frac{1}{2}$-inch plates, and though it may materially increase the unit-strength of thinner plates of *mild* steel, their excess strength is variable, and its amount cannot always be depended on.

18. *Tensile Strength of Plates in the Joint.*—The author has not met with any experiments giving the tensile unit-strength of the net-plate area between the holes in single-riveted lap joints of steel with *punched* holes, but it is probable that the bending of the joint does not injure punched steel plates, unless they are very thick, to the same extent as iron on account of the greater ductility of the steel. Table XXXI. contains experiments by Professor Kennedy on the tensile strength of steel plates in single-riveted lap joints with *drilled* holes, and it will be observed that bending does not seem to injure drilled plates, though we have previously seen that it reduces the shearing strength of the rivets in single-riveted lap joints.

TABLE XXXI.—*Tensile strength of Steel Plates in single-riveted Lap Joints with* DRILLED *Holes, derived from experiments in which the Plate tore* (Kennedy).

Mode of riveting.	Maker and size of Plate.	No. of Rivets and size of Hole.		Pitch.	Tensile strength of solid Plate per square inch.	Tensile strength of Plate in Joints per sq. inch of net area.*	Bearing pressure per square inch.	Authority and Reference.
		No.	Diameter.					
			inch.	inch.	tons.	tons.	tons.	
Hand	Landore, $1\frac{3}{4} \times \frac{3}{8}$ inch	2	0·531	0·875	29·97	**29·94**	19·19	Kennedy, Proc. I. M. E., 1881, p. 248.
,,	,, $2 \times \frac{3}{8}$ inch	2	0·531	1·00	29·97	**29·16**	25·71	,, ,, ,,
,,	,, $3 \times \frac{3}{8}$ inch	2	0·781	1·50	29·97	**34·40**	31·63	,, ,, ,,
,,	,, $11·84 \times \frac{3}{8}$ inch	7	0·86	1·68	29·33	**33·48**	32·45	,, ,, p. 254.
,,	,, $10·45 \times \frac{3}{8}$ inch	7	0·78	1·49	29·33	**36·69**	33·31	,, ,, ,,
,,	,, $4 \times \frac{3}{4}$ inch	2	1·1	2·00	27·27	**29·73**	24·28	,, ,, p. 717.
,,	,, $4·2 \times \frac{3}{4}$ inch	2	1·2	2·15	27·27	**30·25**	22·83	,, ,, ,,
,,	,, $3·6 \times \frac{3}{4}$ inch	2	1·1	1·81	27·27	**33·24**	21·31	,, ,, ,,
,,	,, $4 \times \frac{3}{4}$ inch	2	1·1	2·00	26·93	**29·72**	24·16	,, ,, ,,

* In this and the following tables the term "net area" means the net area of the plate straight across one row of rivet holes, not zigzag.

We also find that *drilling* the holes has materially raised the tensile unit-strength of the plate in all but the first two experiments, the excess in the others averaging more than 10 per cent. over the unit-strength of the solid plate.

Table XXXII. contains experiments made by Kirkaldy, for the Board of Trade and for Mr. Moberly, on double and triple-riveted lap joints with *punched* holes, and they are very valuable on account of the large size of most of the joints. It will be observed that the tensile unit-strength of all the plates not exceeding $\frac{5}{16}$-inch in thickness, and which tore straight or nearly straight across, either equalled or exceeded that of the solid plate, and that the $\frac{7}{16}$-inch plates in which the fracture took a zigzag direction tore at a much lower unit-stress than the original plate, showing that punching had probably injured them and also that the longitudinal distance between the transverse rows of rivets was not sufficient to make them break straight across and thus develop their maximum strength. It will also be observed that plates exceeding $\frac{1}{2}$-inch thick tore at a much lower unit-stress than the original plate in consequence of the injury done by punching, the thicker plates suffering most.

Table XXXIII. contains experiments on double and triple-riveted lap joints with *drilled* holes. In these we again find that the tensile unit-strength of the plate, in consequence of drilling, was much greater than that of the solid plate, the excess averaging more than 10 per cent. in the thin plates, but considerably less in the $\frac{3}{4}$ and 1-inch plates.

Table XXXIV. contains experiments made by Kirkaldy for the Board of Trade on triple-riveted lap joints with plates *punched* and *annealed*, and here we find that the tensile unit-strength of the plate is considerably in excess of that of the original plate, and that this excess diminishes with the thickness of the plate.

TABLE XXXII.—*Tensile strength of Steel Plates in double or triple-riveted Lap Joints with* PUNCHED *Holes, derived from experiments in which the Plate tore.*

Mode of riveting	Maker and size of Plate.	No. of Rivets and size of Hole.		Pitch.	Tensile strength of solid Plate per sq. inch.	Tensile strength of Plate in joints per sq. inch of net area.	Riveting.	Mode of Fracture.	Authority and Reference.
		No.	Diam.						
			inch.	inch.	tons.	tons.			
Machine	Steel Co. of Scotland—								
„	$13\frac{1}{2} \times \frac{1}{4}$ in.	24 (8 rivets in each row).	0·47	$1\frac{11}{16}$	29·9	31·7	Triple-riveted chain riveting.	Tore straight across one row of rivets.	Board of Trade, Merch. Shipping Expts. on Steel, pp. 94 to 97.
„	$13 \times \frac{1}{2}$ in.	21 (7 rivets in each row).	0·66	$1\frac{7}{8}$	28·0	30·3	„	„	„
„	$11\frac{3}{4} \times \frac{3}{4}$ in.	12 (4 rivets in each row).	0·97	2·94	28·7	27·8	„	„	„
„	$12\frac{1}{4} \times 1$ in.	„	1·09	3·06	28·4	13·6	„	„	„
	Landore Steel—								
„	$8\frac{1}{4} \times \frac{7}{16}$ in.	6 (3 rivets in each row),	0·81	2·75	27·9	25·65	Double-riveted zigzag riveting.	Tore through zigzag line of rivets.	Moberly, Proc. Inst. C. E., Vol. LXXII., pp. 246, 247.
„	$8\frac{1}{4} \times \frac{7}{16}$ in.	„	0·81	2·75	28·5	25·8	„	„	„
„	$8\frac{1}{4} \times \frac{7}{16}$ in.	„	0·81	2·75	28·0	26·0	„	Greater part straight.	„
„	$8\frac{1}{4} \times \frac{7}{16}$ in.	„	0·81	2·75	28·8	25·5	„	Straight.	„
„	$11 \times \frac{7}{16}$ in.	8 (4 rivets in each row).	0·81	2·75	28·7	27·0	„	Part straight, part zigzag.	„
„	$7\frac{1}{2} \times \frac{5}{16}$ in.	6 (3 rivets in each row).	0·68	2·50	28·8	29·0	„	Greater part straight.	„
„	$7\frac{1}{2} \times \frac{5}{16}$ in.	„	0·68	2·50	29·3	29·31	„	„	„
„	$10 \times \frac{5}{16}$ in.	8 (4 rivets in each row).	0·68	2·50	29·2	29·2	„	„	„

TABLE XXXIII.—*Tensile strength of Steel Plates in double or triple-riveted Lap Joints with* DRILLED *Holes, derived from experiments in which the Plate tore.*

Mode of riveting.	Maker and size of Plate.	No. of Rivets and size of Hole.		Pitch.	Tensile strength of Solid Plate per sq inch	Tensile strength of Plate in joints per sq. inch of net area.	Bearing pressure per square inch.	Riveting.	Mode of Fracture.	Experimenter.	Authority and Reference.
		No.	Diam.								
			in.	in.	tons.	tons.	tons.				
Machine,	Steel Co. of Scotland— $13\frac{1}{2} \times \frac{1}{4}$ in.	24 (8 rivets in each row).	·45	$1\frac{11}{16}$	31·6	35·96	—	Triple-riveted chain riveting.	Tore straight across one row of rivets.	Kirkaldy.	Board of Trade, Merch. Shipping Expts. on Steel, pp. 82 to 85.
"	$13 \times \frac{1}{2}$ in.	21 (7 rivets in each row).	·64	$1\frac{7}{8}$	29·1	34·18	—	"	"	"	"
"	$11\frac{3}{4} \times \frac{3}{4}$ in.	12 (4 rivets in each row).	·95	$2\frac{15}{16}$	28·6	30·46	—	"	"	"	"
"	$12\frac{1}{2} \times 1$ in.	"	1·08	$3\frac{1}{16}$	27·7	29·44	—	"	"	"	"
Hand,	Landore— $10 \times \frac{3}{8}$ in.	7 (4 rivets in one row, 3 in other).	0·8	2·9	29·97	33·33	40·6	Double-riveted zigzag riveting.	—	Kennedy.	Kennedy, Proc. Inst. M.E., 1885.
Hydraulic,	$10 \times \frac{3}{8}$ in.	"	0·8	2·9	29·97	32·93	40·07	"	—	"	"
Hand,	$11 \times \frac{3}{4}$ in.	"	1·1	3·1	28·46	31·21	26·30	"	—	"	"
Hydraulic,	$11 \times \frac{3}{4}$ in.	"	1·1	3·1	28·46	30·15	25·48	"	—	"	"

TABLE XXXIV.—*Tensile strength of Steel Plates in triple-riveted Lap Joints, with Plates* PUNCHED *and* ANNEALED *derived from experiments in which the Plate tore* (Board of Trade).

Mode of riveting.	Maker and size of Plate.	No. of Rivets and size of Hole.		Pitch.	Tensile strength of solid Plate per sq. inch.	Tensile strength of Plate in joints per sq. inch of net area.	Riveting.	Mode of Fracture.	Authority and Reference.
		No.	Diam.						
			in.	in.	tons.	tons.			
Machine	Steel Co. of Scotland— 13¾ × ¼ in.	24 (8 rivets in each row).	·47	1½	29·9	32·9	Triple-riveted chain.	Tore straight across one row of rivets.	Board of Trade, Merch. Shipping Expts. on Steel, p. 95 to 97.
,,	13 × ½ in.	21 (7 rivets in each row).	·66	1½	28·0	32·2	,,	,,	,,
,,	11¾ × ⅝ in.	12 (4 rivets in each row).	·97	2·94	28·7	30·6	,,	,,	,,
,,	12¼ × 1 in.	12 (4 rivets in each row).	1·09	3·06	28·4	29·5	,,	,,	,,

TABLE XXXV.—*Tensile strength of Steel Plates in double-riveted and double-covered Butt Joints with* DRILLED *Holes, derived from experiments in which the Plate tore.*

Mode of riveting.	Maker and size of Plate.	No. of Rivets and size of Hole.		Pitch.	Tensile strength of Solid Plate per sq. inch.	Tensile strength of Plate in joints per sq. inch of net area.	Bearing pressure per square inch.	Riveting.	Mode of Fracture.	Experimenter.	Authority and Reference.
		No.	Diam.								
			in.	in.	tons.	tons.	tons.				
—	12 × $\frac{11}{16}$ in.	6 (3 rivets in each row).	$1\frac{1}{16}$	4·0	27·0	**24·6**	—	Zigzag riveting.	Part straight, part zigzag.	- -	Martell, Trans. I. N. A., 1878, p. 15.
—	12 × $\frac{11}{16}$ in.	,,	$1\frac{1}{16}$	4·0	27·0	**23·1**	—	,,	,,	—	,,
—	12 × $\frac{11}{16}$ in.	,,	$1\frac{1}{16}$	4·0	27·0	**28·7**	—	Chain riveting.	Tore straight across.	—	,,
Machine,	Landore— 11·32 × $\frac{9}{16}$ in.	,,	0·81	3·77	28·2	**26·7**	—	Zigzag riveting.	Tore through zigzag line.	Kirkaldy	Moberly, Proc. I. C. E. Vol. LXIX., p. 361.
Hand,	9·66 × $\frac{3}{8}$ in.	7	0·7	2·75	28·24	**30·36**	42·28	,,	Tore straight across.	Kennedy	Kennedy, Proc. I. M. E., 1885.
Hydraulic,	9·66 × $\frac{3}{8}$ in.	7	0·7	2·75	28·24	**30·72**	42·33	,,	Mean of 4 Expts., 2 of which tore straight across and 2 nearly so.	,,	,,
,,	9·66 × $\frac{3}{8}$ in.	7	0·7	2·75	28·24	**29·21**	40·24	,,	Tore straight across.	,,	,,
Hand,	13·2 × $\frac{3}{4}$ in.	6	1·1	4·4	27·83	**26·47**	39·49	,,	,,	,,	
Hydraulic,	13·2 × $\frac{3}{4}$ in.	6	1·1	4·4	27·83	**25·07**	37·43	,,	Said to have been riveted at too low a hydraulic pressure.	,,	,,
,,	13·2 × $\frac{3}{4}$ in.	6	1·1	4·4	27·83	**25·13**	37·54	,,	,,	,,	,,

Table XXXV. contains experiments on double-riveted and double-covered butt joints, with *drilled* holes; and it will be observed that in those experiments in which the plates tore straight across, the tensile unit-strength exceeded that of the solid plate, but not at all to the same extent as in lap joints, no doubt because the transverse pitch of double-covered butt joints is much greater than that of lap joints, and the benefit derived from perforation is therefore of smaller amount.

Mr. Longridge states that he has found in chain riveting with iron plates and iron rivets a considerable excess of strength compared to zigzag riveting, both in butt and lap joints, and the subject is worthy of further investigation.

The unit-strengths of mild steel plates in the joint differ, as has been just shown, according as the plates are punched, or drilled, or punched and annealed, and the following table, derived from Kirkaldy's experiments made for the Board of Trade on triple-riveted lap joints (*See* Tables XXXII., XXXIII., and XXXIV.), illustrates this very clearly:*—

TABLE XXXVI.—*Unit-strength of Plate between the holes in triple-riveted Lap Joints compared with that of the solid Plate* (Steel made by Steel Co. of Scotland.) (Board of Trade.)

Specimen.	Unit-strength of net section between the holes compared with that of the solid plate (= 100).			
	¼ inch.	½ inch.	¾ inch.	1 inch.
	per cent.	per cent.	per cent.	per cent.
Punched - - - -	106·0	107·9	96·6	47·8
Punched and annealed -	109·9	114·8	106·5	103·8
Drilled - - - -	113·8	117·2	106·5	105·9

It will be observed that when the thickness of the plate did not exceed ½ inch, the unit-strength of the plate in the joint

* Merchant Shipping Experiments on Steel, p. 29.

with punched holes was greater than that of the solid plate, but that it was 3·4 per cent. less in the ¾-inch plate, and over 50 per cent. less in the 1-inch plate. At first sight there may seem to be a discrepancy between the punched plates in this table and in Table XXVII., which was derived from experiments on perforated plates, made by the same makers, and tested by the same experimenter, and which were punched but not riveted together in a joint. Nevertheless, other experimenters have arrived at the same conclusion—namely, that the unit-strength of the net section of steel plates in *punched* joints is often greater than that of the solid plate as ascertained by test samples and the following is offered as a possible explanation. When experimenting on riveted iron joints for the Royal Irish Academy,[*] the author tested six perforated samples, all taken from the same ⅜-inch iron plate, whose tensile strength in the solid was 24 tons per square inch. Each sample was 8 inches wide and perforated with 5 holes each ·82 inch in diameter. Three of these samples were punched and three drilled, and hot rivets (dummy rivets) were inserted and riveted up in one of the punched and in one of the drilled samples. Their tensile strengths, compared with a test sample of the solid plate, were as follows:—

	Strength per square inch of net section. Tons.
Solid iron plate before perforation	24
Plate punched, but without rivets (mean of 2)	22·1
Plate drilled, but without rivets (mean of 2)	23·8
Plate punched and hot rivets inserted	25·16
Plate drilled and hot rivets inserted	28·84

We see here that the insertion of hot rivets and riveting them up greatly increased the tensile strength of the perforated plates. This result was perhaps partly due to the

[*] Report on Riveted Joints, Trans. R.I.A., Vol. XXV.

contraction of the rivets, which caused their heads and points to grasp and hold together with strong friction the parts on each side of the holes, and possibly in a slight degree to the annealing effect of the hot rivets on the plate, but the increased strength was also probably to some extent due to the rivets preventing the holes from becoming oval under longitudinal stress, for Professor Unwin states that "in an experiment by Mr. Adamson the strength of a perforated plate was increased by driving a pin into the hole, so as to prevent the metal round the hole from collapsing into an elliptical shape." * This explanation, it must be confessed, is not altogether satisfactory, for drilled joints ought, as well as punched joints, to derive benefit from the rivets preventing contraction of the holes, and, as far as the author is aware, this has not been observed by experimenters. Mr. Wildish thinks the recuperative effect of the riveting on the strength of punched steel plates may perhaps be due to the bending of the joint, and that with such a ductile material as mild steel this bending may have the opposite effect that it has with iron and may reduce the injury done by the punch to mild steel.† On the other hand, Mr. Traill and his assistants attributed the great amount of injury sustained by the 1-inch punched joint (Table XXXII.) to the bending of the joint, which, however, probably affects a very thick punched plate differently from a thin one.‡

The experiments described in this Section prove that the metal between the rivet holes, when steel plates are drilled, or when they are punched and annealed, has generally a considerably higher tensile unit-strength than the original plate, the excess being greater with thin than with thick plates; and Professor Kennedy found in his experiments that "this excess

* Proc. Inst. M.E., 1881, p. 325.
† Trans. Inst. Nav. Arch., 1885, p. 184.
‡ Merch. Ship. Expts., p. 30.

tenacity (with Landore SS steel) amounted to more than 20 per cent. (both in ⅜-inch and ¾-inch plates) when the pitch of the rivets was about 1·9 diameters. In other cases ⅜-inch plate gave an excess of 15 per cent. at fracture with a pitch of two diameters, of 10 per cent. with a pitch of 3·6 diameters, and of 6·6 per cent. with a pitch of 3·9 diameters; and ¾-inch plate 7·8 per cent. excess with a pitch of 2·8 diameters." *

If the shearing strength of steel rivets, and the tensile strength of steel plates between the holes in riveted joints average 23 tons and 30 tons respectively, the sectional area of the rivets should be nearly one-third greater than the net area of the plate, in this respect differing from iron riveting where the rivet area and the net plate area are very nearly equal to each other.

19. *Margin and Lap of Plates.*—Prof. Kennedy found that drilled steel plates required the same margin as in iron joints, that is, that a margin (or net distance from outside of holes to edge of plate) equal to the diameter of the hole, was sufficient for steel, but he qualifies this in the case of high bearing pressure (as in double-covered butts), in which case he thinks it would probably be wise to increase the margin; and this view is corroborated by Mr. Moberly's experiments on double-covered butt joints, in several of which, especially those with punched holes, the margin of the plate burst out in front of the rivet—in one instance in the exceedingly rare manner of a piece of the margin being shorn out for the full width of the rivet. † It will be observed that several of the double-riveted joints in Tables XXXII. and XXXV. broke along the zigzag line, and when this occurs the plate evidently tears at a lower stress than it would if the longitudinal pitch (distance longitudinally between the centres of the transverse rows of rivets) were greater, so as to insure its tearing straight

---

* Proc. Inst. M. E., 1885.
† Proc. Inst. C.E., Vol. LXIX., plate 8.

across. To attain, therefore, the maximum strength of a joint the lap must be sufficiently wide to prevent the plate tearing along a zigzag line. The stress along this zigzag line is neither a purely tensile nor a purely shearing stress, but a compound of both; and Mr. Moberly's experiments show that the capacity of mild steel plates, $\frac{5}{16}$ and $\frac{7}{16}$ inch thick, to resist this compound stress is from 75 to 80 per cent. of their tensile strength; accordingly the net section along the zigzag line should be about $\frac{1}{3}$rd greater than that along the straight row of rivets. On this subject Prof. Kennedy observes:—" It has been found (with drilled holes) that the net metal measured zigzag should be from 30 to 35 per cent. in excess of that measured straight across, in order to insure a straight fracture. This corresponds to a diagonal pitch of $\frac{2}{3}p + \frac{d}{3}$, if p be the straight pitch and d the diameter of the rivet-hole. To find the proper breadth of lap for a double-riveted joint, it is probably best to proceed by first setting this pitch off, and then finding from it the longitudinal pitch, or distance between the centres of the lines of rivets."[*]

20. *Friction of Steel Joints and Slip of Plates.*—The friction of two plates held together by a single rivet is theoretically nearly equal to the contractile strength of the rivet multiplied by the coefficient of friction of one plate pressing on another, and in a joint it is theoretically proportional to the sectional area of the rivets in the joint. From experiments made by the author, the coefficient of friction for ordinary steel plates pressing on each other is about 0·6, and it has been shown in § 12 that the contractile strength of hand-made (nominal) $\frac{3}{4}$-inch iron rivets does not exceed 7 tons per rivet, or 12·32 tons per square inch of rivet area, as this stress will pull the heads off. Consequently, the theoretic friction of two steel plates held together by a $\frac{3}{4}$-inch

[*] Proc. Inst. M. E., 1885.

iron rivet cannot much exceed $0.6 \times 7 = 4.2$ tons, or say 7·4 tons per square inch of rivet area. Sir E. Reed describes some experiments in which one plate was single-riveted with 3 rivets in a row between two other plates, so that the rivets were in double shear, and both sides of the middle plate were thus subject to friction against the covers. With iron plates and iron rivets slipping took place at a mean pull of 4·6 and 5·6 tons per rivet for ¾-inch and 1-inch rivets respectively. With corresponding steel plates and steel rivets slipping took place at mean pulls of 4·1 and 5·5 tons respectively for ¾-inch and 1-inch rivets, or with very slightly less friction than with the iron.* Mr. Wildish describes later experiments with steel rivets and steel plates made at Pembroke Dockyard, the middle plate being riveted as before with 3 rivets in a row between two other plates, so that the rivets were again in double shear. The 1-inch rivets were used in ⅜-inch steel plates and the ¾-inch rivets in ½-inch plates. The following mean frictional resistances per rivet were obtained with hand riveting:—

	1-in. rivet. tons.	¾-in. rivet. tons.
With snap head and point,	6·4	4·72
With pan head and boiler point,	7·36	4·52
With pan head and countersunk point,	8·55	6·25
With countersunk head and point,	9·04	4·95

These results are not in proportion to the sectional area of the rivets, but they indicate that on the whole the friction is greatest for the countersunk rivets, in this respect differing from the earlier Admiralty experiments, in which countersunk rivets caused less friction than rivets with pan heads and conical points (see § 10). Some other experiments were made with machine riveting with snap heads and points, and the mean friction per rivet was 9·6 tons for the 1-inch

* Wilson on steam boilers, p. 58.

rivets and 5·9 tons for the ¾-inch rivets, or a good deal higher than for the corresponding "hand" rivets with snap heads and points, but very little superior to "hand" rivets with countersunk heads and points.*

It might be supposed that the friction per rivet of double-covered butt joints would be nearly twice as great as that of corresponding lap joints, but this does not appear to have been the case in Professor Kennedy's experiments, in several of which the friction per rivet in the butt joints scarcely exceeded that in the corresponding lap joints, and in others it was about one-third greater, and he gives the following approximate table, derived from his experiments with steel rivets and drilled steel plates, of the loads per rivet at which a joint will commence to slip visibly, and he adds the following rule:—

"To find the probable load at which a joint of any breadth will commence to slip, it is only necessary to multiply the number of rivets in the given breadth by the proper figure taken from the last column of the table. It will be understood that the figures are not given as exact."†

TABLE XXXVII.—*Approximate Loads per rivet at which a joint will commence to slip visibly* (Kennedy).

Rivet diameter.	Type of joint.	Riveting.	Slipping load per rivet.
inch.			tons.
¾	Single-riveted,	Hand,	2·5
,,	Double-riveted,	,,	3·0 to 3·5
,,	,,	Machine,	7
1	Single-riveted,	Hand,	3·2
,,	Double-riveted,	,,	4·3
,,	,,	Machine,	8 to 10

* Wildish, Trans. Inst. Nav. Arch., 1885, p. 190.
† Proc. Inst. M. E., 1885.

It will be observed in Professor Kennedy's experiments that the friction with machine riveting was twice as great as with hand riveting, and this friction, as has been already mentioned in § 10, is an important factor in the staunchness of boilers, the working and proof stresses of which should, whenever practicable, not exceed the stress at which incipient slipping of the joint occurs.

21. *Bearing Pressure of Rivets.*—In Professor Kennedy's experiments on single-riveted lap joints with bearing pressures of 50 to 55 tons per square inch the rivets sheared in most cases at stresses varying from 16 to 18 tons per square inch, and he concludes that the bearing pressure in lap joints should probably not exceed 42 to 43 tons per square inch, and that in double-covered butt joints a pressure of from 45 to 50 tons will cause shearing to take place at from 16 to 18 tons per square inch. High bearing pressure also seemed to have prevented the full excess tenacity of the drilled plate (of 10 to 12 per cent.) from being reached, although it did not actually weaken the plate below its normal resistance.[*] Mr. Moberly states that in his experiments "the bearing pressure per square inch of surface on the rivets varied a good deal, and it does not appear to have borne any relation to the shearing of the rivets, or to the yielding of the joint in any other way. As far as these experiments go, it appears that it may reach 50 tons per square inch at the moment of fracture."[†] Perhaps we may adopt provisionally as standard *crippling pressures* for steel rivets—40 tons per square inch for lap joints, and 50 tons for the middle plate of double-covered butt joints, inasmuch as the bearing pressure of the rivet against the middle plate in the butt is evidently more uniformly distributed than it is in a lap joint which bends under severe stress.

---

[*] Proc. Inst. M. E. 1885.
[†] Proc. Inst. C. E. Vol. LXXII., p. 241.

22. *Efficiency of Steel Joints*.—The following table gives the estimated efficiency of steel joints, deduced chiefly from the experiments described in the previous pages, and which were made by Kirkaldy for the Board of Trade and for Mr. Moberly, and by Professor Kennedy for the Research Committee of the Institution of Mechanical Engineers. Calling the total tensile strength of the original solid plate 100, the approximate efficiency of various joints are as follows, provided the pitch of the riveting is such that the joint is on the point of giving way from the tearing of the plates or the shearing of the rivets indifferently, and provided that the punched plates are as mild and ductile as those made by the Landore Siemens Steel Company, and provided that in double riveting the longitudinal pitch is sufficient to prevent the plates tearing along a zigzag line.

TABLE XXXVIII.—*Approximate Efficiency of steel joints of various kinds compared with that of the solid plate* (= 100).

	Efficiency per cent.		
	Thickness of plates.		
	inch. $\frac{1}{4}$ to $\frac{3}{8}$	inch. $\frac{1}{2}$ to $\frac{5}{8}$	inch. $\frac{3}{4}$ to $\frac{7}{8}$
Original solid plate, - - - -	100	100	100
Lap joint, single-riveted, punched, - -	50	45	40
Do. do. drilled, - -	55	50	45
Do. double-riveted, punched, - -	75	70	65
Do. do. drilled, -	80	75	70
Butt joint, double-covered, double-riveted, punched,	75	70	65
Do. do. do. drilled, -	80	75	70

Future experiments may modify the foregoing percentages, and it will be recollected that extra soft rivet steel will seriously reduce the efficiency of a joint whose proportions are based on the supposition that its shearing strength is up to the usual standard, and that in the commoner kinds of steel, such as those used in girderwork and shipbuilding, the excess strength due to drilling is less likely to be attained than with boiler plates, and the efficiency of joints with such plates will probably be (say 10 per cent.) less than those given in the table.

23. *Proportions of Joints—Boilermakers' Practice—Ship-builders' Practice—Girderwork.*—The author is indebted to Mr. Moberly for the following tables of the proportions he adopts for the riveting of steel boilers at the Engineering Works of Messrs. Easton and Anderson at Erith, the holes being *punched* and the plates *not annealed* after punching, and he has added in each table the efficiency he calculates for the joints thus proportioned.

Let d = diameter of rivet,
d_1 = diameter of punch,
t = thickness of plate,
p = pitch of rivets transversely,
m = pitch of rivets longitudinally, that is, distance apart of transverse pitch lines in double-riveted joints,
L = lap of plates,
$d_1 + \dfrac{t}{5}$ = diameter of die.

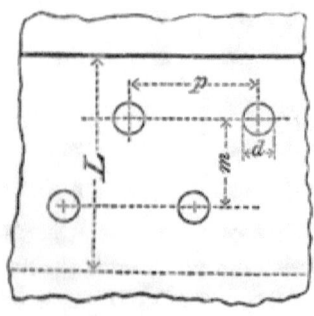

TABLE XXXIX.—*Proportions of single-riveted lap joints for steel boilers* (Moberly).

—	1st series for maximum strength.					2nd series with $d = 2t$ (generally).				
Thickness of plates.	d	d_1	p	L	Calculated efficiency.	d	d_1	p	L	Calculated efficiency.
inch.	in.	inch.	inch.	inch.	per cent.	inch.	inch.	inch.	inch.	per cent.
$\frac{3}{16}$	$\frac{3}{16}$	.47	$1\frac{1}{8}$	$1\frac{3}{16}$	60·0	$\frac{3}{8}$	·4	$\frac{7}{8}$	1	55·5
$\frac{1}{4}$	$\frac{5}{16}$	·61	$1\frac{7}{16}$	$1\frac{1}{2}$	59·3	$\frac{1}{2}$	·54	$1\frac{3}{16}$	$1\frac{7}{16}$	55·9
$\frac{5}{16}$	$\frac{3}{8}$	·81	2	$2\frac{1}{16}$	60·0	$\frac{5}{8}$	·68	$1\frac{1}{2}$	$1\frac{5}{8}$	54·4
$\frac{3}{8}$	$\frac{7}{8}$	·94	$2\frac{5}{16}$	$2\frac{3}{8}$	59·5	$\frac{3}{4}$	·81	$1\frac{13}{16}$	2	54·8
$\frac{7}{16}$	1	1·06	$2\frac{5}{8}$	$2\frac{2}{3}$	58·0	$\frac{7}{8}$	·94	$2\frac{3}{16}$	$2\frac{7}{16}$	54·8
$\frac{1}{2}$	$1\frac{1}{8}$	1·19	3	3	56·7	1	1·06	$2\frac{1}{2}$	$2\frac{5}{8}$	53·7
$\frac{9}{16}$	$1\frac{1}{4}$	1·315	$3\frac{5}{16}$	$3\frac{3}{8}$	53·5	$1\frac{1}{16}$	1·12	$2\frac{9}{16}$	$2\frac{3}{4}$	50·6

N.B.—1st series should be generally used, except for circular seams of boilers with horizontal double-riveted lap joints, *for which use 2nd series.*

TABLE XL.—*Proportions of double-riveted lap joints for steel boilers* (Moberly).

Thickness of plates.	d	d_1	p	m	L	Calculated efficiency.
inch.	inch.	inch.	inch.	inch.	inch.	per cent.
$\frac{3}{16}$	$\frac{3}{8}$	·4	$1\frac{1}{2}$	$\frac{7}{8}$	$2\frac{1}{2}$	74·4
$\frac{1}{4}$	$\frac{1}{2}$	·54	2	$1\frac{3}{16}$	$2\frac{7}{8}$	76·0
$\frac{5}{16}$	$\frac{5}{8}$	·68	$2\frac{1}{2}$	$1\frac{1}{2}$	$3\frac{1}{4}$	74·0
$\frac{3}{8}$	$\frac{3}{4}$	·81	3	$1\frac{3}{4}$	$4\frac{1}{5}$	74·4
$\frac{7}{16}$	$\frac{7}{8}$	·94	$3\frac{1}{2}$	2	$4\frac{2}{3}$	71·8
$\frac{1}{2}$	1	1·06	4	$2\frac{3}{16}$	$5\frac{1}{2}$	69·8
$\frac{9}{16}$	$1\frac{1}{16}$	1·12	$4\frac{1}{4}$	$2\frac{1}{2}$	$5\frac{7}{8}$	67·4

OF RIVETED JOINTS.

TABLE XLI.—*Proportions of double-riveted and double-covered butt joints for steel boilers* (Moberly).

Thickness of plates.	Thickness of covers.		d	d_1	p	m	L	Calculated efficiency.
	Outer.	Inner.						
inch.	inch.	inch.	inch.	inch.	inch.	inch.	inch.	per cent.
$\frac{3}{8}$	$\frac{5}{16}$	$\frac{1}{4}$	$\frac{5}{8}$	·68	3	$1\frac{3}{4}$	$4\frac{1}{4}$	78·5
$\frac{7}{16}$	$\frac{5}{16}$	$\frac{5}{16}$	$\frac{11}{16}$	·74	$3\frac{1}{2}$	2	5	77·7
$\frac{1}{2}$	$\frac{3}{8}$	$\frac{3}{8}$	$\frac{3}{4}$	·81	4	$2\frac{3}{16}$	$5\frac{1}{2}$	75·9
$\frac{9}{16}$	$\frac{7}{16}$	$\frac{7}{16}$	$\frac{13}{16}$	·87	$4\frac{1}{2}$	$2\frac{3}{8}$	$6\frac{3}{8}$	74·0

On these tables Mr. Moberly makes the following observations:—" The tables of proportions here given are calculated from data furnished by experiments actually made with $\frac{9}{16}''$ double-riveted, double-covered, butt joints, in 1881,* and with $\frac{7}{16}''$ and $\frac{5}{16}''$ double-riveted lap joints, in 1882.† The plates were of Landore S quality, with a tensile strength of 30 tons per square inch, and the strength of the plates in the joints, between the rivet holes, along the straight line is taken—

For thickness of plate, - $\frac{3}{16}''$, $\frac{1}{4}''$, $\frac{5}{16}''$, $\frac{3}{8}''$, $\frac{7}{16}''$, $\frac{1}{2}''$, $\frac{9}{16}''$,
Strength per sq. inch in tons, 32, 32, 31, 31, 30, 29, 28.

The shearing strength of the rivets has been taken at 23 tons per square inch for single-riveted lap joints, and at 25 tons per square inch for all others. The diameter of the die is one-fifth the thickness of the plate greater than the diameter of the punch—*i.e.*, diameter of die $= d_1 + \frac{t}{5}$. As the strength of the plate *in the joint* varies greatly with its thickness, the above experiments do not warrant any conclusion as to the strength of plates over $\frac{9}{16}''$ thick, and the tables have, therefore,

* Proc. Inst. C. E., Vol. LXIX.
† *Idem*, Vol. LXXII.

not been carried above that thickness. If the holes are punched small and rimered, or drilled out of the solid, the strength of the plate in the joint is increased, especially with the thicker plates, but it is uncertain as yet to what extent. It would probably have the effect of raising the efficiency of all the double-riveted lap joints to 75 per cent., and that of *all* the double-riveted, double-covered butt joints to 80 per cent., and might therefore be worth doing for the thicker plates, but not for the thinner ones. As the rivets have an excess of strength in most cases, especially in the thicker plates, the proportions would probably not require altering for holes thus punched." It will be observed that in punching steel Mr. Moberly makes the diameter of the die equal to that of the punch, plus 20 per cent. of the thickness of the plate, whereas in punching iron the usual allowance is only 12·5 per cent. (p. 20). In punching the holes for the Board of Trade's experiments, already described, the dies were larger than the punches by about one-fifth of the diameter, this proportion being adopted as affording a moderate degree of clearance to the punch.* In the discussion on Professor Kennedy's paper on riveted joints, at the Institution of Mechanical Engineers, in 1885, Mr. Moberly said "that in order to get good riveted joints the rivet must be properly put in, and must fill the hole when cold, but as a matter of fact this is really never the case in consequence of the rivets contracting laterally in cooling. This is not so much the case with steel rivets, as they are, or should be, worked at a dull red heat. The practice of his firm had been originally to work the rivets at a higher temperature, but they found a great advantage in not going above a dull red. The most perfect joint would be one in which drilled holes were used and the rivets turned and closed quite cold." †

* Merch. Ship. Expts. on Steel, p. 18.
† Engineering, May, 1885, p. 524.

Mr. F. W. Webb, Locomotive Superintendent of the London and North Western Railway, describes his method of dealing with steel boiler plates as follows:—

"He first punched all the holes by a template machine, using a large bolster; he then raised the plates just to a blood-red heat, and put them on one side to cool; and then did all the other work cold. When the plates were cold, they were bent into the proper circle, sponged with sal-ammoniac, and put together. After the boilers were made, the whole of the scale was taken off inside with sal-ammoniac and water; and as soon as they were dry they were given a coating of glycerine, if the boiler was likely not to be used at once, in order that corrosion might not go any further. The small ends of the two punched holes were put together; and he liked to have a very deep head for the rivets, so that they should not give way. Rivet heads were sometimes so flat that they actually curled up round the edge, and so caused leakage. In the longitudinal joints he put the rivets in double shear by using a thin cover-plate outside and inside. In that case, and that only, he punched the hole about $\frac{1}{8}$ smaller than the finished size, and then put a rimer through the three thicknesses together, and they came as true as possible. On this method he was making boilers at the rate of four every week—locomotive boilers working at 140 lbs. pressure, and they gave no trouble whatever." Mr. Webb finds that sponging the two inside surfaces of a joint with hot water and sal-ammoniac, before putting together, eats off the magnetic (black) oxide, which, if left on, gets pounded to powder in the act of riveting, and is thus left in the joint, causing leakage and corrosion. He also finds that this process saves in a great measure the necessity of caulking, as his boilers are practically tight at 160 lbs. pressure when they come out of the shop, without anything further being done.[*] Mr. Webb,

[*] Proc. Inst. M. E., 1881, pp. 261, 263.

after trying various proportions of rivets and distances apart, "had arrived at a single-riveted butt joint with double cover-plate, the dimensions being, for a $\frac{7}{16}$-inch plate, rivets $\frac{3}{4}$ inch diameter, 2-inch pitch, distance from edge of plate about 1½ inch, cover-plates ⅜ inch thick, 5¼ inch wide." The efficiency of this joint was stated to be 71·6 per cent. of the strength of the solid plate.*

Mr. Wildish describes the Admiralty practice in the use of steel for shipbuilding in the following terms:—" With regard to the treatment of the steel plates in working them into the hull, it may be well to remark that as long ago as 1878 it was decided not to anneal them after punching, as a means of making good the injury to the material due to the punching. All the butt straps, however, to the plating forming an important feature in the general structural strength, such as the outside plating, deck plating, stringers, &c., were ordered to have the holes drilled in them, or to be annealed after the holes were punched. The countersunk holes in the outside plating were at the same time ordered to be punched about ⅛th of an inch smaller in diameter than the full size required, the enlargement of the holes being made in countersinking, which was to be carried through the whole thickness. Countersunk riveting was also to be adopted in the stringers, deck-plating, and other parts subject to considerable tensile stress, the holes being treated similarly to those in the outside plating. This practice is now (1885) followed in the service, and the annealing, both of the frame bars as well as the plates, is only carried out in very exceptional instances, and in order to relieve the tensions set up in the material in bending it to shape."† Mr. Baker states that " in proportioning the riveted joints of the tubes and other members (of the Forth Bridge), the shearing area is generally made one and a-half

---

* Proc. Inst. M. E., 1879, p. 304.
† Trans. Inst. Nav. Arch., 1885, p. 181.

times the net sectional area of the plates connected if in tension, and half that for planed and butted joints in compression only." The specified minimum strengths of the steel used for compression and tension members are 34 and 30 tons respectively, and the shearing strength of the rivet steel varies from 22 to 24 tons per square inch.*

24. *Theoretic Proportions of Steel Joints.*—The theoretic proportions of steel joints may be found in the manner already described in § 14, p. 41, and, using the same symbols as before, we have the following rules:—

a. To find the theoretic diameter of a steel rivet in single shear, we have from equation 3 (p. 42),

$$\frac{d}{t} = \frac{c}{\cdot 7854\, s} \qquad (11)$$

Example 1. What is the ratio $\frac{d}{t}$ in a single-riveted lap joint with *drilled* holes?

Here, $c = 40$ tons per square inch (see § 21),
           $s = 22$ tons per square inch (see p. 55).

Answer. $\frac{d}{t} = \frac{40}{\cdot 7854 \times 22} = 2\cdot 31$

With ½-inch plates, for instance, $d = 1\cdot 16$ inch.

Example 2. What is the ratio $\frac{d}{t}$ in a double-riveted lap joint with *drilled* holes?

Here, $c = 40$ tons per square inch (see § 21),
           $s = 23$ tons per square inch (see p. 55).

Answer. $\frac{d}{t} = \frac{40}{\cdot 7854 \times 23} = 2\cdot 2$

With ½-inch plates, for instance, $d = 1\cdot 1$ inch.

b. To find the theoretic diameter of a steel rivet in double shear, the shearing strength of a steel rivet in double shear being *twice* that in single shear (p. 54), we have from equation 11,

$$\frac{d}{t} = \frac{c}{2 \times \cdot 7854\, s} = \frac{c}{1\cdot 57\, s} \qquad (12)$$

* Engineering, Sept., 1884, p. 224.

Example 3. What is the ratio $\frac{d}{t}$ in double-covered butt joints, either single or double-riveted, with *drilled* holes?

Here, $c = 50$ tons per square inch (see § 21),

$s = 23$ tons per square inch (see p. 55).

Answer. $\dfrac{d}{t} = \dfrac{50}{1\cdot 57 \times 23} = 1\cdot 38$

With ½-inch plates, for instance, $d = \cdot 7$ inch.

c. To find the theoretic pitch in single-riveted joints, either in single or double shear. From equation 5 (p. 42),

$$\frac{p}{d} = \frac{c+f}{f} \qquad (13)$$

Example 4. What is the ratio $\frac{p}{d}$ in a single-riveted lap joint with *drilled* holes, the tensile unit-strength of the solid plate being 28 tons per square inch, but increased by drilling to 30 tons in the perforated plate?

Here, $c = 40$ tons per square inch (see § 21),

$f = 30$ tons per square inch.

Answer. $\dfrac{p}{d} = \dfrac{40 + 30}{30} = 2\cdot 33$

With ½-inch plates, for instance (see ex. 1), $p = 2\cdot 33 \times 1\cdot 16 = 2\cdot 7$ inches.

Example 5. What is the ratio $\frac{p}{d}$ in a single-riveted and double-covered butt joint with *drilled* holes, the tensile unit-strength of the solid plate being 28 tons per square inch, but increased by drilling to 30 tons in the perforated plate (p. 72)?

Here, $c = 50$ tons per square inch (see § 21),

$f = 30$ tons per square inch.

Answer. $\dfrac{p}{d} = \dfrac{50 + 30}{30} = 2\cdot 67$

With ½-inch plates, for instance (see ex. 3), $p = 2\cdot 67 \times \cdot 7 = 1\cdot 87$ inches.

d. To find the theoretic pitch in double-riveted joints, either in single or double shear. From equation 8 (p. 43),

$$\frac{p}{d} = \frac{2c+f}{f} \qquad (14)$$

Example 6. What is the ratio $\frac{p}{d}$ in a double-riveted lap joint with *drilled* holes, the tensile unit-strength of the solid plate being 28 tons per square inch, but increased by drilling to 29 tons in the perforated plate, the excess being less than in the two previous examples in consequence of the increased pitch (p. 72)?

Here, $c = 40$ tons per square inch (see § 21),
$f = 29$ tons per square inch.

Answer. $\dfrac{p}{d} = \dfrac{80 + 29}{29} = 3\cdot76$

With ½-inch plates, for instance (see ex. 2), $p = 3\cdot76 \times 1\cdot1 = 4\cdot14$ inches.

Example 7. What is the ratio $\dfrac{p}{d}$ in a double-riveted and double-covered butt joint with *drilled* holes, the tensile strength of the solid plate being 28 tons per square inch, but increased by drilling to 29 tons in the perforated plate?

Here, $c = 50$ tons per square inch (see § 21),
$f = 29$ tons per square inch.

Answer. $\dfrac{p}{d} = \dfrac{100 + 29}{29} = 4\cdot45$

With ⅓-inch plates, for instance (see ex. 3), $p = 4\cdot45 \times \cdot7 = 3\cdot12$ inches.

1892.

BOOKS RELATING

TO

APPLIED SCIENCE

PUBLISHED BY

E. & F. N. SPON,

LONDON: 125, STRAND.

NEW YORK: 12, CORTLANDT STREET.

The Engineers' Sketch-Book of Mechanical Movements, Devices, Appliances, Contrivances, Details employed in the Design and Construction of Machinery for every purpose. Collected from numerous Sources and from Actual Work. Classified and Arranged for Reference. *Nearly* 2000 *Illustrations.* By T. B. BARBER, Engineer. Second Edition, 8vo, cloth, 7s. 6d.

A Pocket-Book for Chemists, Chemical Manufacturers, Metallurgists, Dyers, Distillers, Brewers, Sugar Refiners, Photographers, Students, etc., etc. By THOMAS BAYLEY, Assoc. R.C. Sc. Ireland, Analytical and Consulting Chemist and Assayer. Fifth edition, 481 pp., royal 32mo, roan, gilt edges, 5s.

SYNOPSIS OF CONTENTS:

Atomic Weights and Factors—Useful Data—Chemical Calculations—Rules for Indirect Analysis—Weights and Measures—Thermometers and Barometers—Chemical Physics—Boiling Points, etc.—Solubility of Substances—Methods of Obtaining Specific Gravity—Conversion of Hydrometers—Strength of Solutions by Specific Gravity—Analysis—Gas Analysis—Water Analysis—Qualitative Analysis and Reactions—Volumetric Analysis—Manipulation—Mineralogy—Assaying—Alcohol—Beer—Sugar—Miscellaneous Technological matter relating to Potash, Soda, Sulphuric Acid, Chlorine, Tar Products, Petroleum, Milk, Tallow, Photography, Prices, Wages, Appendix, etc., etc.

The Mechanician: A Treatise on the Construction and Manipulation of Tools, for the use and instruction of Young Engineers and Scientific Amateurs, comprising the Arts of Blacksmithing and Forging; the Construction and Manufacture of Hand Tools, and the various Methods of Using and Grinding them; description of Hand and Machine Processes; Turning and Screw Cutting. By CAMERON KNIGHT, Engineer. *Containing* 1147 *illustrations,* and 397 pages of letter-press. Fourth edition, 4to, cloth, 18s.

B

Just Published, in Demy 8vo, cloth, containing 975 pages and 250 Illustrations, price 7s. 6d.

SPONS' HOUSEHOLD MANUAL:
A Treasury of Domestic Receipts and Guide for Home Management.

PRINCIPAL CONTENTS.

Hints for selecting a good House, pointing out the essential requirements for a good house as to the Site, Soil, Trees, Aspect, Construction, and General Arrangement; with instructions for Reducing Echoes, Waterproofing Damp Walls, Curing Damp Cellars.

Sanitation.—What should constitute a good Sanitary Arrangement; Examples (with Illustrations) of Well- and Ill-drained Houses; How to Test Drains; Ventilating Pipes, etc.

Water Supply.—Care of Cisterns; Sources of Supply; Pipes; Pumps; Purification and Filtration of Water.

Ventilation and Warming.—Methods of Ventilating without causing cold draughts, by various means; Principles of Warming; Health Questions; Combustion; Open Grates; Open Stoves; Fuel Economisers, Varieties of Grates; Close-Fire Stoves; Hot-air Furnaces; Gas Heating; Oil Stoves; Steam Heating; Chemical Heaters; Management of Flues; and Cure of Smoky Chimneys.

Lighting.—The best methods of Lighting; Candles, Oil Lamps, Gas, Incandescent Gas, Electric Light; How to test Gas Pipes; Management of Gas.

Furniture and Decoration.—Hints on the Selection of Furniture; on the most approved methods of Modern Decoration; on the best methods of arranging Bells and Calls; How to Construct an Electric Bell.

Thieves and Fire.—Precautions against Thieves and Fire; Methods of Detection; Domestic Fire Escapes; Fireproofing Clothes, etc.

The Larder.—Keeping Food fresh for a limited time; Storing Food without change, such as Fruits, Vegetables, Eggs, Honey, etc.

Curing Foods for lengthened Preservation, as Smoking, Salting, Canning, Potting, Pickling, Bottling Fruits, etc.; Jams, Jellies, Marmalade, etc.

The Dairy.—The Building and Fitting of Dairies in the most approved modern style; Butter-making; Cheesemaking and Curing.

The Cellar.—Building and Fitting; Cleaning Casks and Bottles; Corks and Corking; Aërated Drinks; Syrups for Drinks; Beers; Bitters; Cordials and Liqueurs; Wines; Miscellaneous Drinks.

The Pantry.—Bread-making; Ovens and Pyrometers; Yeast; German Yeast; Biscuits; Cakes; Fancy Breads; Buns.

The Kitchen.—On Fitting Kitchens; a description of the best Cooking Ranges, close and open; the Management and Care of Hot Plates, Baking Ovens, Dampers, Flues, and Chimneys; Cooking by Gas; Cooking by Oil; the Arts of Roasting, Grilling, Boiling, Stewing, Braising, Frying.

Receipts for Dishes—Soups, Fish, Meat, Game, Poultry, Vegetables, Salads, Puddings, Pastry, Confectionery, Ices, etc., etc.; Foreign Dishes.

The Housewife's Room.—Testing Air, Water, and Foods; Cleaning and Renovating; Destroying Vermin.

Housekeeping, Marketing.

The Dining-Room.—Dietetics; Laying and Waiting at Table; Carving; Dinners, Breakfasts, Luncheons, Teas, Suppers, etc.

The Drawing-Room.—Etiquette; Dancing; Amateur Theatricals; Tricks and Illusions; Games (indoor).

The Bedroom and Dressing-Room; Sleep; the Toilet; Dress; Buying Clothes; Outfits; Fancy Dress.

The Nursery.—The Room; Clothing; Washing; Exercise; Sleep; Feeding; Teething; Illness; Home Training.

The Sick-Room.—The Room; the Nurse; the Bed; Sick Room Accessories; Feeding Patients; Invalid Dishes and Drinks; Administering Physic; Domestic Remedies; Accidents and Emergencies; Bandaging; Burns; Carrying Injured Persons; Wounds; Drowning; Fits; Frost-bites; Poisons and Antidotes; Sunstroke; Common Complaints; Disinfection, etc.

The Bath-Room.—Bathing in General; Management of Hot-Water System.
The Laundry.—Small Domestic Washing Machines, and methods of getting up linen; Fitting up and Working a Steam Laundry.
The School-Room.—The Room and its Fittings; Teaching, etc.
The Playground.—Air and Exercise; Training; Outdoor Games and Sports.
The Workroom.—Darning, Patching, and Mending Garments.
The Library.—Care of Books.
The Garden.—Calendar of Operations for Lawn, Flower Garden, and Kitchen Garden.
The Farmyard.—Management of the Horse, Cow, Pig, Poultry, Bees, etc., etc.
Small Motors.—A description of the various small Engines useful for domestic purposes, from 1 man to 1 horse power, worked by various methods, such as Electric Engines, Gas Engines, Petroleum Engines, Steam Engines, Condensing Engines, Water Power, Wind Power, and the various methods of working and managing them.
Household Law.—The Law relating to Landlords and Tenants, Lodgers, Servants, Parochial Authorities, Juries, Insurance, Nuisance, etc.

On Designing Belt Gearing. By E. J. COWLING WELCH, Mem. Inst. Mech. Engineers, Author of 'Designing Valve Gearing.' Fcap. 8vo, sewed, 6d.

A Handbook of Formulæ, Tables, and Memoranda, for Architectural Surveyors and others engaged in Building. By J. T. HURST, C.E. Fourteenth edition, royal 32mo, roan, 5s.

"It is no disparagement to the many excellent publications we refer to, to say that in our opinion this little pocket-book of Hurst's is the very best of them all, without any exception. It would be useless to attempt a recapitulation of the contents, for it appears to contain almost everything that anyone connected with building could require, and, best of all, made up in a compact form for carrying in the pocket, measuring only 5 in. by 3 in., and about ¾ in. thick, in a limp cover. We congratulate the author on the success of his laborious and practically compiled little book, which has received unqualified and deserved praise from every professional person to whom we have shown it."—*The Dublin Builder.*

Tabulated Weights of Angle, Tee, Bulb, Round, Square, and Flat Iron and Steel, and other information for the use of Naval Architects and Shipbuilders. By C. H. JORDAN, M.I.N.A. Fourth edition, 32mo, cloth, 2s. 6d.

A Complete Set of Contract Documents for a Country Lodge, comprising Drawings, Specifications, Dimensions (for quantities), Abstracts, Bill of Quantities, Form of Tender and Contract, with Notes by J. LEANING, printed in facsimile of the original documents, on single sheets fcap., in paper case, 10s.

A Practical Treatise on Heat, as applied to the Useful Arts; for the Use of Engineers, Architects, &c. By THOMAS BOX. With 14 plates. Sixth edition, crown 8vo, cloth, 12s. 6d.

A Descriptive Treatise on Mathematical Drawing Instruments: their construction, uses, qualities, selection, preservation, and suggestions for improvements, with hints upon Drawing and Colouring. By W. F. STANLEY, M.R.I. Sixth edition, *with numerous illustrations,* crown 8vo, cloth, 5s.

Quantity Surveying. By J. LEANING. With 42 illustrations. Second edition, revised, crown 8vo, cloth, 9s.

CONTENTS :

A complete Explanation of the London Practice.
General Instructions.
Order of Taking Off.
Modes of Measurement of the various Trades.
Use and Waste.
Ventilation and Warming.
Credits, with various Examples of Treatment.
Abbreviations.
Squaring the Dimensions.
Abstracting, with Examples in illustration of each Trade.
Billing.
Examples of Preambles to each Trade.
Form for a Bill of Quantities.
    Do.   Bill of Credits.
    Do.   Bill for Alternative Estimate.
Restorations and Repairs, and Form of Bill.
Variations before Acceptance of Tender.
Errors in a Builder's Estimate.

Schedule of Prices.
Form of Schedule of **Prices**.
Analysis of Schedule of **Prices**.
Adjustment of Accounts.
Form of a Bill of Variations.
Remarks on Specifications.
Prices and Valuation of Work, with Examples and Remarks upon each Trade.
The Law as it affects Quantity Surveyors, with Law Reports.
Taking Off after the Old Method.
Northern Practice.
The General Statement of the Methods recommended by the Manchester Society of Architects for taking Quantities.
Examples of Collections.
Examples of " Taking Off" in each Trade.
Remarks on the Past and Present Methods of Estimating.

Spons' Architects' and Builders' Price Book, with useful Memoranda. Edited by W. YOUNG, Architect. Crown 8vo, cloth, red edges, 3s. 6d. *Published annually.* Nineteenth edition. *Now ready.*

Long-Span Railway Bridges, comprising Investigations of the Comparative Theoretical and Practical Advantages of the various adopted or proposed Type Systems of Construction, with numerous Formulæ and Tables giving the weight of Iron or Steel required in Bridges from 300 feet to the limiting Spans; to which are added similar Investigations and Tables relating to Short-span Railway Bridges. Second and revised edition. By B. BAKER, Assoc. Inst. C.E. *Plates*, crown 8vo, cloth, 5s.

Elementary Theory and Calculation of Iron Bridges and Roofs. By AUGUST RITTER, Ph.D., Professor at the Polytechnic School at Aix-la-Chapelle. Translated from the third German edition, by H. R. SANKEY, Capt. R.E. With 500 *illustrations*, 8vo, cloth, 15s.

The Elementary Principles of Carpentry. By THOMAS TREDGOLD. Revised from the original edition, and partly re-written, by JOHN THOMAS HURST. Contained in 517 pages of letter-press, and *illustrated with 48 plates and* 150 *wood engravings*. Sixth edition, reprinted from the third, crown 8vo, cloth, 12s. 6d.

Section I. On the Equality and Distribution of Forces—Section II. Resistance of Timber—Section III. Construction of Floors—Section IV. Construction of Roofs—Section V. Construction of Domes and Cupolas—Section VI. Construction of Partitions—Section VII. Scaffolds, Staging, and Gantries—Section VIII. Construction of Centres for Bridges—Section IX. Coffer-dams, Shoring, and Strutting—Section X. Wooden Bridges and Viaducts—Section XI. Joints, Straps, and other Fastenings—Section XII. Timber.

The Builder's Clerk : a Guide to the Management of a Builder's Business. By THOMAS BALES. Fcap. 8vo, cloth, 1s. 6d.

Practical Gold-Mining: a Comprehensive Treatise on the Origin and Occurrence of Gold-bearing Gravels, Rocks and Ores, and the methods by which the Gold is extracted. By C. G. WARNFORD LOCK, co-Author of 'Gold: its Occurrence and Extraction.' With 8 plates and 275 engravings in the text, royal 8vo, cloth, 2l. 2s.

Hot Water Supply: A Practical Treatise upon the Fitting of Circulating Apparatus in connection with Kitchen Range and other Boilers, to supply Hot Water for Domestic and General Purposes. With a Chapter upon Estimating. *Fully illustrated*, crown 8vo, cloth, 3s.

Hot Water Apparatus: An Elementary Guide for the Fitting and Fixing of Boilers and Apparatus for the Circulation of Hot Water for Heating and for Domestic Supply, and containing a Chapter upon Boilers and Fittings for Steam Cooking. 32 *illustrations*, fcap. 8vo, cloth, 1s. 6d.

The Use and Misuse, and the Proper and Improper Fixing of a Cooking Range. Illustrated, fcap. 8vo, sewed, 6d.

Iron Roofs: Examples of Design, Description. *Illustrated with* 64 *Working Drawings of Executed Roofs.* By ARTHUR T. WALMISLEY, Assoc. Mem. Inst. C.E. Second edition, revised, imp. 4to, half-morocco, 3l. 3s.

A History of Electric Telegraphy, to the Year 1837. Chiefly compiled from Original Sources, and hitherto Unpublished Documents, by J. J. FAHIE, Mem. Soc. of Tel. Engineers, and of the International Society of Electricians, Paris. Crown 8vo, cloth, 9s.

Spons' Information for Colonial Engineers. Edited by J. T. HURST. Demy 8vo, sewed.

No. 1, Ceylon. By ABRAHAM DEANE, C.E. 2s. 6d.

CONTENTS:

Introductory Remarks—Natural Productions—Architecture and Engineering—Topography, Trade, and Natural History—Principal Stations—Weights and Measures, etc., etc.

No. 2. Southern Africa, including the Cape Colony, Natal, and the Dutch Republics. By HENRY HALL, F.R.G.S., F.R.C.I. With Map. 3s. 6d.

CONTENTS:

General Description of South Africa—Physical Geography with reference to Engineering Operations—Notes on Labour and Material in Cape Colony—Geological Notes on Rock Formation in South Africa—Engineering Instruments for Use in South Africa—Principal Public Works in Cape Colony: Railways, Mountain Roads and Passes, Harbour Works, Bridges, Gas Works, Irrigation and Water Supply, Lighthouses, Drainage and Sanitary Engineering, Public Buildings, Mines—Table of Woods in South Africa—Animals used for Draught Purposes—Statistical Notes—Table of Distances—Rates of Carriage, etc.

No. 3. India. By F. C. DANVERS, Assoc. Inst. C.E. With Map. 4s. 6d.

CONTENTS:

Physical Geography of India—Building Materials—Roads—Railways—Bridges—Irrigation—River Works—Harbours—Lighthouse Buildings—Native Labour—The Principal Trees of India—Money—Weights and Measures—Glossary of Indian Terms, etc.

Our Factories, Workshops, and Warehouses: their Sanitary and Fire-Resisting Arrangements. By B. H. THWAITE, Assoc. Mem. Inst. C.E. With 183 *wood engravings*, crown 8vo, cloth, 9s.

A Practical Treatise on Coal Mining. By GEORGE G. ANDRÉ, F.G.S., Assoc. Inst. C.E., Member of the Society of Engineers. With 82 *lithographic plates*. 2 vols., royal 4to, cloth, 3l. 12s.

A Practical Treatise on Casting and Founding, including descriptions of the modern machinery employed in the art. By N. E. SPRETSON, Engineer. Fifth edition, with 82 *plates* drawn to scale, 412 pp., demy 8vo, cloth, 18s.

A Handbook of Electrical Testing. By H. R. KEMPE, M.S.T.E. Fourth edition, revised and enlarged, crown 8vo, cloth, 16s.

The Clerk of Works: a Vade-Mecum for all engaged in the Superintendence of Building Operations. By G. G. HOSKINS, F.R.I.B.A. Third edition, fcap. 8vo, cloth, 1s. 6d.

American Foundry Practice: Treating of Loam, Dry Sand, and Green Sand Moulding, and containing a Practical Treatise upon the Management of Cupolas, and the Melting of Iron. By T. D. WEST, Practical Iron Moulder and Foundry Foreman. Second edition, *with numerous illustrations*, crown 8vo, cloth, 10s. 6d.

The Maintenance of Macadamised Roads. By T. CODRINGTON, M.I.C.E., F.G.S., General Superintendent of County Roads for South Wales. Second edition, 8vo, cloth, 7s. 6d.

Hydraulic Steam and Hand Power Lifting and Pressing Machinery. By FREDERICK COLYER, M. Inst. C.E., M. Inst. M.E. With 73 *plates*, 8vo, cloth, 18s.

Pumps and Pumping Machinery. By F. COLYER, M.I.C.E., M.I.M.E. With 23 *folding plates*, 8vo, cloth, 12s. 6d.

Pumps and Pumping Machinery. By F. COLYER. Second Part. With 11 *large plates*, 8vo, cloth, 12s. 6d.

A Treatise on the Origin, Progress, Prevention, and Cure of Dry Rot in Timber; with Remarks on the Means of Preserving Wood from Destruction by Sea-Worms, Beetles, Ants, etc. By THOMAS ALLEN BRITTON, late Surveyor to the Metropolitan Board of Works, etc., etc. With 10 *plates*, crown 8vo, cloth, 7s. 6d.

The Artillery of the Future and the New Powders. By J. A. LONGRIDGE, Mem. Inst. C.E. 8vo, cloth, 5s.

Gas Works: their Arrangement, Construction, Plant, and Machinery. By F. COLYER, M. Inst. C.E. *With 31 folding plates,* 8vo, cloth, 12s. 6d.

The Municipal and Sanitary Engineer's Handbook. By H. PERCY BOULNOIS, Mem. Inst. C.E., Borough Engineer, Portsmouth. *With numerous illustrations.* Second edition, demy 8vo, cloth, 15s.

CONTENTS:

The Appointment and Duties of the Town Surveyor—Traffic—Macadamised Roadways—Steam Rolling—Road Metal and Breaking—Pitched Pavements—Asphalte—Wood Pavements—Footpaths—Kerbs and Gutters—Street Naming and Numbering—Street Lighting—Sewerage—Ventilation of Sewers—Disposal of Sewage—House Drainage—Disinfection—Gas and Water Companies, etc., Breaking up Streets—Improvement of Private Streets—Borrowing Powers—Artizans' and Labourers' Dwellings—Public Conveniences—Scavenging, including Street Cleansing—Watering and the Removing of Snow—Planting Street Trees—Deposit of Plans—Dangerous Buildings—Hoardings—Obstructions—Improving Street Lines—Cellar Openings—Public Pleasure Grounds—Cemeteries—Mortuaries—Cattle and Ordinary Markets—Public Slaughter-houses, etc.—Giving numerous Forms of Notices, Specifications, and General Information upon these and other subjects of great importance to Municipal Engineers and others engaged in Sanitary Work.

Metrical Tables. By Sir G. L. MOLESWORTH, M.I.C.E. 32mo, cloth, 1s. 6d.

CONTENTS.

General—Linear Measures—Square Measures—Cubic Measures—Measures of Capacity—Weights—Combinations—Thermometers.

Elements of Construction for Electro-Magnets. By Count TH. DU MONCEL, Mem. de l'Institut de France. Translated from the French by C. J. WHARTON. Crown 8vo, cloth, 4s. 6d.

A Treatise on the Use of Belting for the Transmission of Power. By J. H. COOPER. Second edition, *illustrated,* 8vo, cloth, 15s.

A Pocket-Book of Useful Formulæ and Memoranda for Civil and Mechanical Engineers. By Sir GUILFORD L. MOLESWORTH, Mem. Inst. C.E. *With numerous illustrations,* 744 pp. Twenty-second edition, 32mo, roan, 6s.

SYNOPSIS OF CONTENTS:

Surveying, Levelling, etc.—Strength and Weight of Materials—Earthwork, Brickwork, Masonry, Arches, etc.—Struts, Columns, Beams, and Trusses—Flooring, Roofing, and Roof Trusses—Girders, Bridges, etc.—Railways and Roads—Hydraulic Formulæ—Canals, Sewers, Waterworks, Docks—Irrigation and Breakwaters—Gas, Ventilation, and Warming—Heat, Light, Colour, and Sound—Gravity: Centres, Forces, and Powers—Millwork, Teeth of Wheels, Shafting, etc.—Workshop Recipes—Sundry Machinery—Animal Power—Steam and the Steam Engine—Water-power, Water-wheels, Turbines, etc.—Wind and Windmills—Steam Navigation, Ship Building, Tonnage, etc.—Gunnery, Projectiles, etc.—Weights, Measures, and Money—Trigonometry, Conic Sections, and Curves—Telegraphy—Mensuration—Tables of Areas and Circumference, and Arcs of Circles—Logarithms, Square and Cube Roots, Powers—Reciprocals, etc.—Useful Numbers—Differential and Integral Calculus—Algebraic Signs—Telegraphic Construction and Formulæ.

Hints on Architectural Draughtsmanship. By G. W. TUXFORD HALLATT. Fcap. 8vo, cloth, 1s. 6d.

Spons' Tables and Memoranda for Engineers; selected and arranged by J. T. HURST, C.E., Author of 'Architectural Surveyors' Handbook,' 'Hurst's Tredgold's Carpentry,' etc. Eleventh edition, 64mo, roan, gilt edges, 1s.; or in cloth case, 1s. 6d.

This work is printed in a pearl type, and is so small, measuring only 2½ in. by 1¾ in. by ¼ in. thick, that it may be easily carried in the waistcoat pocket.

"It is certainly an extremely rare thing for a reviewer to be called upon to notice a volume measuring but 2½ in. by 1¾ in., yet these dimensions faithfully represent the size of the handy little book before us. The volume—which contains 118 printed pages, besides a few blank pages for memoranda—is, in fact, a true pocket-book, adapted for being carried in the waist-coat pocket, and containing a far greater amount and variety of information than most people would imagine could be compressed into so small a space. . . . . The little volume has been compiled with considerable care and judgment, and we can cordially recommend it to our readers as a useful little pocket companion."—*Engineering.*

A Practical Treatise on Natural and Artificial Concrete, its Varieties and Constructive Adaptations. By HENRY REID, Author of the 'Science and Art of the Manufacture of Portland Cement.' New Edition, *with* 59 *woodcuts and* 5 *plates,* 8vo, cloth, 15s.

Notes on Concrete and Works in Concrete; especially written to assist those engaged upon Public Works. By JOHN NEWMAN, Assoc. Mem. Inst. C.E., crown 8vo, cloth, 4s. 6d.

Electricity as a Motive Power. By Count TH. DU MONCEL, Membre de l'Institut de France, and FRANK GERALDY, Ingénieur des Ponts et Chaussées. Translated and Edited, with Additions, by C. J. WHARTON, Assoc. Soc. Tel. Eng. and Elec. *With* 113 *engravings and diagrams,* crown 8vo, cloth, 7s. 6d.

Treatise on Valve-Gears, with special consideration of the Link-Motions of Locomotive Engines. By Dr. GUSTAV ZEUNER, Professor of Applied Mechanics at the Confederated Polytechnikum of Zurich. Translated from the Fourth German Edition, by Professor J. F. KLEIN, Lehigh University, Bethlehem, Pa. *Illustrated,* 8vo, cloth, 12s. 6d.

The French-Polisher's Manual. By a French-Polisher; containing Timber Staining, Washing, Matching, Improving, Painting, Imitations, Directions for Staining, Sizing, Embodying, Smoothing, Spirit Varnishing, French-Polishing, Directions for Re-polishing. Third edition, royal 32mo, sewed, 6d.

Hops, their Cultivation, Commerce, and Uses in various Countries. By P. L. SIMMONDS. Crown 8vo, cloth, 4s. 6d.

The Principles of Graphic Statics. By GEORGE SYDENHAM CLARKE, Major Royal Engineers. *With* 112 *illustrations.* Second edition, 4to, cloth, 12s. 6d.

Dynamo Tenders' Hand-Book. By F. B. BADT, late 1st Lieut. Royal Prussian Artillery. *With 70 illustrations.* Third edition, 18mo, cloth, 4*s.* 6*d.*

Practical Geometry, Perspective, and Engineering Drawing; a Course of Descriptive Geometry adapted to the Requirements of the Engineering Draughtsman, including the determination of cast shadows and Isometric Projection, each chapter being followed by numerous examples; to which are added rules for Shading, Shade-lining, etc., together with practical instructions as to the Lining, Colouring, Printing, and general treatment of Engineering Drawings, with a chapter on drawing Instruments. By GEORGE S. CLARKE, Capt. R.E. Second edition, *with 21 plates.* 2 vols., cloth, 10*s.* 6*d.*

The Elements of Graphic Statics. By Professor KARL VON OTT, translated from the German by G. S. CLARKE, Capt. R.E., Instructor in Mechanical Drawing, Royal Indian Engineering College. *With 93 illustrations,* crown 8vo, cloth, 5*s.*

A Practical Treatise on the Manufacture and Distribution of Coal Gas. By WILLIAM RICHARDS. Demy 4to, with *numerous wood engravings and 29 plates,* cloth, 28*s.*

SYNOPSIS OF CONTENTS:

Introduction—History of Gas Lighting—Chemistry of Gas Manufacture, by Lewis Thompson, Esq., M.R.C.S.—Coal, with Analyses, by J Paterson, Lewis Thompson, and G. R. Hislop, Esqrs.—Retorts, Iron and Clay—Retort Setting—Hydraulic Main—Condensers—Exhausters—Washers and Scrubbers—Purifiers—Purification—History of Gas Holder—Tanks, Brick and Stone, Composite, Concrete, Cast-iron, Compound Annular Wrought-iron—Specifications—Gas Holders—Station Meter—Governor—Distribution—Mains—Gas Mathematics, or Formulæ for the Distribution of Gas, by Lewis Thompson, Esq.—Services—Consumers' Meters—Regulators—Burners—Fittings—Photometer—Carburization of Gas—Air Gas and Water Gas—Composition of Coal Gas, by Lewis Thompson, Esq.—Analyses of Gas—Influence of Atmospheric Pressure and Temperature on Gas—Residual Products—Appendix—Description of Retort Settings, Buildings, etc., etc.

The New Formula for Mean Velocity of Discharge of Rivers and Canals. By W. R. KUTTER. Translated from articles in the 'Cultur-Ingénieur,' by LOWIS D'A. JACKSON, Assoc. Inst. C.E. 8vo, cloth, 12*s.* 6*d.*

The Practical Millwright and Engineer's Ready Reckoner; or Tables for finding the diameter and power of cog-wheels, diameter, weight, and power of shafts, diameter and strength of bolts, etc. By THOMAS DIXON. Fourth edition, 12mo, cloth, 3*s.*

Tin: Describing the Chief Methods of Mining, Dressing and Smelting it abroad; with Notes upon Arsenic, Bismuth and Wolfram. By ARTHUR G. CHARLETON, Mem. American Inst. of Mining Engineers. *With plates,* 8vo, cloth, 12*s.* 6*d.*

Perspective, Explained and Illustrated. **By G. S.** CLARKE, Capt. R.E. *With illustrations*, 8vo, cloth, **3s. 6d.**

Practical Hydraulics; a Series of Rules and Tables for the use of Engineers, etc., etc. By THOMAS BOX. Ninth edition, *numerous plates*, post 8vo, cloth, **5s.**

The Essential Elements of Practical Mechanics; based on the Principle of Work, designed for Engineering Students. By OLIVER BYRNE, formerly Professor of Mathematics, College for Civil Engineers. Third edition, *with* **148** *wood engravings*, post 8vo, cloth, **7s. 6d.**

CONTENTS:

Chap. 1. How Work is Measured by a Unit, both with and without reference to a Unit of Time—Chap. 2. The Work of Living Agents, the Influence of Friction, and introduces one of the most beautiful Laws of Motion—Chap. 3. The principles expounded in the first and second chapters are applied to the Motion of Bodies—Chap. 4. The Transmission of Work by simple Machines—Chap. 5. Useful Propositions and Rules.

Breweries and Maltings: their Arrangement, Construction, Machinery, and Plant. By G. SCAMELL, F.R.I.B.A. Second edition, revised, enlarged, and partly rewritten. By F. COLYER, M.I.C.E., M.I.M.E. *With* **20** *plates*, 8vo, cloth, **12s. 6d.**

A Practical Treatise on the Construction of Horizontal and Vertical Waterwheels, specially designed for the use of operative mechanics. By WILLIAM CULLEN, Millwright and Engineer. *With* **11** *plates*. Second edition, revised and enlarged, small 4to, cloth, **12s. 6d.**

A Practical Treatise on Mill-gearing, **Wheels, Shafts,** *Riggers, etc.;* for the use of Engineers. By THOMAS BOX. Third edition, *with* **11** *plates*. Crown 8vo, cloth, **7s. 6d.**

Mining Machinery: a Descriptive Treatise on the Machinery, Tools, and other Appliances used in Mining. By G. G. ANDRÉ, F.G.S., Assoc. Inst. C.E., Mem. of the Society of Engineers. Royal 4to, uniform with the Author's Treatise on Coal Mining, containing **182** *plates*, accurately drawn to scale, with descriptive text, in **2 vols.**, cloth, **3l. 12s.**

CONTENTS:

Machinery for Prospecting, Excavating, Hauling, and Hoisting—Ventilation—Pumping—**Treatment of** Mineral Products, including Gold and Silver, Copper, Tin, and Lead, Iron, Coal, Sulphur, China Clay, Brick Earth, etc.

Tables for Setting out Curves for Railways, Canals, Roads, etc., varying from a radius of five chains to three miles. By A. KENNEDY and R. W. HACKWOOD. *Illustrated* 32mo, cloth, **2s. 6d.**

***Practical Electrical Notes and Definitions** for the use of Engineering Students and Practical Men.* By W. PERREN MAYCOCK, Assoc. M. Inst. E.E., Instructor in Electrical Engineering at the Pitlake Institute, Croydon, together with the Rules and Regulations to be observed in Electrical Installation Work. Second edition. Royal 32mo, roan, gilt edges, 4s. 6d., or cloth, red edges, 3s.

***The Draughtsman's Handbook of Plan and Map** Drawing;* including instructions for the preparation of Engineering, Architectural, and Mechanical Drawings. *With numerous illustrations in the text, and 33 plates* (15 *printed in colours*). By G. G. ANDRÉ, F.G.S., Assoc. Inst. C.E. 4to, cloth, 9s.

CONTENTS:

The Drawing Office and its Furnishings—Geometrical Problems—Lines, Dots, and their Combinations—Colours, Shading, Lettering, Bordering, and North Points—Scales—Plotting—Civil Engineers' and Surveyors' Plans—Map Drawing—Mechanical and Architectural Drawing—Copying and Reducing Trigonometrical Formulæ, etc., etc.

***The Boiler-maker's** and Iron Ship-builder's Companion,* comprising a series of original and carefully calculated tables, of the utmost utility to persons interested in the iron trades. By JAMES FODEN, author of 'Mechanical Tables,' etc. Second edition revised, *with illustrations,* crown 8vo, cloth, 5s.

Rock Blasting: a Practical Treatise on the means employed in Blasting Rocks for Industrial Purposes. By G. G. ANDRÉ, F.G.S., Assoc. Inst. C.E. *With 56 illustrations and 12 plates,* 8vo, cloth, 10s. 6d.

Experimental Science: Elementary, Practical, and Experimental Physics. By GEO. M. HOPKINS. *Illustrated by 672 engravings.* In one large vol., 8vo, cloth, 15s.

***A Treatise on Ropemaking** as practised in public and private Rope-yards,* with a Description of the Manufacture, Rules, Tables of Weights, etc., adapted to the Trade, Shipping, Mining, Railways, Builders, etc. By R. CHAPMAN, formerly foreman to Messrs. Huddart and Co., Limehouse, and late Master Ropemaker to H.M. Dockyard, Deptford. Second edition, 12mo, cloth, 3s.

Laxton's Builders' and Contractors' Tables; for the use of Engineers, Architects, Surveyors, Builders, Land Agents, and others. Bricklayer, containing 22 tables, with nearly 30,000 calculations. 4to, cloth, 5s.

Laxton's Builders' and Contractors' Tables. Excavator, Earth, Land, Water, and Gas, containing 53 tables, with nearly 24,000 calculations. 4to, cloth, 5s.

Egyptian Irrigation. By W. WILLCOCKS, M.I.C.E., Indian Public Works Department, Inspector of Irrigation, Egypt. With Introduction by Lieut.-Col. J. C. ROSS, R.E., Inspector-General of Irrigation. *With numerous lithographs and wood engravings*, royal 8vo, cloth, 1*l*. 16*s*.

Screw Cutting Tables for Engineers and Machinists, giving the values of the different trains of Wheels required to produce Screws of any pitch, calculated by Lord Lindsay, M.P., F.R.S., F.R.A.S., etc. Cloth, oblong, 2*s*.

Screw Cutting Tables, for the use of Mechanical Engineers, showing the proper arrangement of Wheels for cutting the Threads of Screws of any required pitch, with a Table for making the Universal Gas-pipe Threads and Taps. By W. A. MARTIN, Engineer. Second edition, oblong, cloth, 1*s*., or sewed, 6*d*.

A Treatise on a Practical Method of Designing Slide-Valve Gears by Simple Geometrical Construction, based upon the principles enunciated in Euclid's Elements, and comprising the various forms of Plain Slide-Valve and Expansion Gearing; together with Stephenson's, Gooch's, and Allan's Link-Motions, as applied either to reversing or to variable expansion combinations. By EDWARD J. COWLING WELCH, Memb. Inst. Mechanical Engineers. Crown 8vo, cloth, 6*s*.

Cleaning and Scouring: a Manual for Dyers, Laundresses, and for Domestic Use. By S. CHRISTOPHER. 18mo, sewed, 6*d*.

A Glossary of Terms used in Coal Mining. By WILLIAM STUKELEY GRESLEY, Assoc. Mem. Inst. C.E., F.G.S., Member of the North of England Institute of Mining Engineers. *Illustrated with numerous woodcuts and diagrams*, crown 8vo, cloth, 5*s*.

A Pocket-Book for Boiler Makers and Steam Users, comprising a variety of useful information for Employer and Workman, Government Inspectors, Board of Trade Surveyors, Engineers in charge of Works and Slips, Foremen of Manufactories, and the general Steam-using Public. By MAURICE JOHN SEXTON. Second edition, royal 32mo, roan, gilt edges, 5*s*.

Electrolysis: a Practical Treatise on Nickeling, Coppering, Gilding, Silvering, the Refining of Metals, and the treatment of Ores by means of Electricity. By HIPPOLYTE FONTAINE, translated from the French by J. A. BERLY, C.E., Assoc. S.T.E. *With engravings.* 8vo, cloth, 9*s*.

Barlow's Tables of Squares, Cubes, Square Roots, Cube Roots, Reciprocals of all Integer Numbers up to 10,000. Post 8vo, cloth, 6s.

A Practical Treatise on the Steam Engine, containing Plans and Arrangements of Details for Fixed Steam Engines, with Essays on the Principles involved in Design and Construction. By ARTHUR RIGG, Engineer, Member of the Society of Engineers and of the Royal Institution of Great Britain. Demy 4to, *copiously illustrated with woodcuts and* 96 *plates*, in one Volume, half-bound morocco, 2l. 2s.; or cheaper edition, cloth, 25s.

This work is not, in any sense, an elementary treatise, or history of the steam engine, but is intended to describe examples of Fixed Steam Engines without entering into the wide domain of locomotive or marine practice. To this end illustrations will be given of the most recent arrangements of Horizontal, Vertical, Beam, Pumping, Winding, Portable, Semi-portable, Corliss, Allen, Compound, and other similar Engines, by the most eminent Firms in Great Britain and America. The laws relating to the action and precautions to be observed in the construction of the various details, such as Cylinders, Pistons, Piston-rods, Connecting-rods, Cross-heads, Motion-blocks, Eccentrics, Simple, Expansion, Balanced, and Equilibrium Slide-valves, and Valve-gearing will be minutely dealt with. In this connection will be found articles upon the Velocity of Reciprocating Parts and the Mode of Applying the Indicator, Heat and Expansion of Steam Governors, and the like. It is the writer's desire to draw illustrations from every possible source, and give only those rules that present practice deems correct.

A Practical Treatise on the Science of Land and Engineering Surveying, Levelling, Estimating Quantities, etc., with a general description of the several Instruments required for Surveying, Levelling, Plotting, etc. By H. S. MERRETT. Fourth edition, revised by G. W. USILL, Assoc. Mem. Inst. C.E. 41 *plates, with illustrations and tables*, royal 8vo, cloth, 12s. 6d.

PRINCIPAL CONTENTS:

Part 1. Introduction and the Principles of Geometry. Part 2. Land Surveying; comprising General Observations—The Chain—Offsets Surveying by the Chain only—Surveying Hilly Ground—To Survey an Estate or Parish by the Chain only—Surveying with the Theodolite—Mining and Town Surveying—Railroad Surveying—Mapping—Division and Laying out of Land—Observations on Enclosures—Plane Trigonometry. Part 3. Levelling—Simple and Compound Levelling—The Level Book—Parliamentary Plan and Section—Levelling with a Theodolite—Gradients—Wooden Curves—To Lay out a Railway Curve—Setting out Widths. Part 4. Calculating Quantities generally for Estimates—Cuttings and Embankments—Tunnels—Brickwork—Ironwork—Timber Measuring. Part 5. Description and Use of Instruments in Surveying and Plotting—The Improved Dumpy Level—Troughton's Level—The Prismatic Compass—Proportional Compass—Box Sextant—Vernier—Pantagraph—Merrett's Improved Quadrant—Improved Computation Scale—The Diagonal Scale—Straight Edge and Sector. Part 6. Logarithms of Numbers—Logarithmic Sines and Co-Sines, Tangents and Co-Tangents—Natural Sines and Co-Sines—Tables for Earthwork, for Setting out Curves, and for various Calculations, etc., etc., etc.

Mechanical Graphics. A Second Course of Mechanical Drawing. With Preface by Prof. PERRY, B.Sc., F.R.S. Arranged for use in Technical and Science and Art Institutes, Schools and Colleges, by GEORGE HALLIDAY, Whitworth Scholar. 8vo, cloth, 6s.

The Assayer's Manual: an Abridged Treatise on the Docimastic Examination of Ores and Furnace and other Artificial Products. By BRUNO KERL. Translated by W. T. BRANNT. *With 65 illustrations*, 8vo, cloth, 12s. 6d.

Dynamo-Electric Machinery: a Text-Book for Students of Electro-Technology. By SILVANUS P. THOMPSON, B.A., D.Sc., M.S.T.E. [*New edition in the press.*

The Practice of Hand Turning in Wood, Ivory, Shell, etc., with Instructions for Turning such Work in Metal as may be required in the Practice of Turning in Wood, Ivory, etc.; also an Appendix on Ornamental Turning. (A book for beginners.) By FRANCIS CAMPIN. Third edition, *with wood engravings*, crown 8vo, cloth, 6s.

CONTENTS:

On Lathes—Turning Tools—Turning Wood—Drilling—Screw Cutting—Miscellaneous Apparatus and Processes—Turning Particular Forms—Staining—Polishing—Spinning Metals—Materials—Ornamental Turning, etc.

Treatise on Watchwork, Past and Present. By the Rev. H. L. NELTHROPP, M.A., F.S.A. *With 32 illustrations,* crown 8vo, cloth, 6s. 6d.

CONTENTS:

Definitions of Words and Terms used in Watchwork—Tools—Time—Historical Summary—On Calculations of the Numbers for Wheels and Pinions; their Proportional Sizes, Trains, etc.—Of Dial Wheels, or Motion Work—Length of Time of Going without Winding up—The Verge—The Horizontal—The Duplex—The Lever—The Chronometer—Repeating Watches—Keyless Watches—The Pendulum, or Spiral Spring—Compensation—Jewelling of Pivot Holes—Clerkenwell—Fallacies of the Trade—Incapacity of Workmen—How to Choose and Use a Watch, etc.

Algebra Self-Taught. By W. P. HIGGS, M.A., D.Sc., LL.D., Assoc. Inst. C.E., Author of 'A Handbook of the Differential Calculus,' etc. Second edition, crown 8vo, cloth, 2s. 6d.

CONTENTS:

Symbols and the Signs of Operation—The Equation and the Unknown Quantity—Positive and Negative Quantities—Multiplication—Involution—Exponents—Negative Exponents—Roots, and the Use of Exponents as Logarithms—Logarithms—Tables of Logarithms and Proportionate Parts—Transformation of System of Logarithms—Common Uses of Common Logarithms—Compound Multiplication and the Binomial Theorem—Division, Fractions, and Ratio—Continued Proportion—The Series and the Summation of the Series—Limit of Series—Square and Cube Roots—Equations—List of Formulæ, etc.

Spons' Dictionary of Engineering, Civil, Mechanical, Military, and Naval; with technical terms in French, German, Italian, and Spanish, 3100 pp., and *nearly 8000 engravings*, in super-royal 8vo, in 8 divisions, 5l. 8s. Complete in 3 vols., cloth, 5l. 5s. Bound in a superior manner, half-morocco, top edge gilt, 3 vols., 6l. 12s.

Notes in Mechanical Engineering. Compiled principally for the use of the Students attending the Classes on this subject at the City of London College. By HENRY ADAMS, Mem. Inst. M.E., Mem. Inst. C.E., Mem. Soc. of Engineers. Crown 8vo, cloth, 2s. 6d.

Canoe and Boat Building: a complete Manual for Amateurs, containing plain and comprehensive directions for the construction of Canoes, Rowing and Sailing Boats, and Hunting Craft. By W. P. STEPHENS. *With numerous illustrations and* 24 *plates of Working Drawings.* Crown 8vo, cloth, 9s.

Proceedings of the National Conference of Electricians, *Philadelphia,* October 8th to 13th, 1884. 18mo, cloth, 3s.

Dynamo-Electricity, its Generation, Application, Transmission, Storage, and Measurement. By G. B. PRESCOTT. *With* 545 *illustrations.* 8vo, cloth, 1l. 1s.

Domestic Electricity for Amateurs. Translated from the French of E. HOSPITALIER, Editor of "L'Electricien," by C. J. WHARTON, Assoc. Soc. Tel. Eng. *Numerous illustrations.* Demy 8vo, cloth, 6s.

CONTENTS:

1. Production of the Electric Current—2. Electric Bells—3. Automatic Alarms—4. Domestic Telephones—5. Electric Clocks—6. Electric Lighters—7. Domestic Electric Lighting—8. Domestic Application of the Electric Light—9. Electric Motors—10. Electrical Locomotion—11. Electrotyping, **Plating,** and Gilding—12. Electric Recreations—13. Various applications—Workshop of the **Electrician.**

Wrinkles in Electric Lighting. By VINCENT STEPHEN. *With illustrations.* 18mo, cloth, 2s. 6d.

CONTENTS:

1. The Electric Current and its production by Chemical means—2. **Production of Electric Currents by Mechanical means**—3. Dynamo-Electric Machines—4. **Electric Lamps**—5. Lead—6. Ship Lighting.

Foundations and Foundation Walls for all classes of Buildings, Pile Driving, Building Stones and Bricks, Pier and Wall construction, Mortars, **Limes,** Cements, Concretes, Stuccos, &c. 64 *illustrations.* By G. T. POWELL and F. BAUMAN. 8vo, cloth, 10s. 6d.

Manual for Gas Engineering Students. By D. LEE. 18mo, cloth, 1s.

Telephones, their Construction and Management. By F. C. Allsop. Crown 8vo, cloth, 5s.

Hydraulic Machinery, Past and Present. A Lecture delivered to the London and Suburban Railway Officials' Association. By H. Adams, Mem. Inst. C.E. *Folding plate.* 8vo, sewed, 1s.

Twenty Years with the Indicator. By Thomas Pray, Jun., C.E., M.E., Member of the American Society of Civil Engineers. 2 vols., royal 8vo, cloth, 12s. 6d.

Annual Statistical Report of the Secretary to the Members of the Iron and Steel Association on the Home and Foreign Iron and Steel Industries in 1889. Issued June 1890. 8vo, sewed, 5s.

Bad Drains, and How to Test them; with Notes on the Ventilation of Sewers, Drains, and Sanitary Fittings, and the Origin and Transmission of Zymotic Disease. By R. Harris Reeves. Crown 8vo, cloth, 3s. 6d.

Well Sinking. The modern practice of Sinking and Boring Wells, with geological considerations and examples of Wells. By Ernest Spon, Assoc. Mem. Inst. C.E., Mem. Soc. Eng., and of the Franklin Inst., etc. Second edition, revised and enlarged. Crown 8vo, cloth, 10s. 6d.

The Voltaic Accumulator: an Elementary Treatise. By Émile Reynier. Translated by J. A. Berly, Assoc. Inst. E.E. *With 62 illustrations,* 8vo, cloth, 9s.

Ten Years' Experience in Works of Intermittent Downward Filtration. By J. Bailey Denton, Mem. Inst. C.E. Second edition, with additions. Royal 8vo, cloth, 5s.

Land Surveying on the Meridian and Perpendicular System. By William Penman, C.E. 8vo, cloth, 8s. 6d.

The Electromagnet and Electromagnetic Mechanism. By Silvanus P. Thompson, D.Sc., F.R.S. 8vo, cloth, 15s.

Incandescent Wiring Hand-Book. By F. B. BADT, late 1st Lieut. Royal Prussian Artillery. *With 41 illustrations and 5 tables.* 18mo, cloth, 4s. 6d.

A Pocket-book for Pharmacists, Medical Practitioners, Students, etc., etc. (British, Colonial, and American). By THOMAS BAYLEY, Assoc. R. Coll. of Science, Consulting Chemist, Analyst, and Assayer, Author of a 'Pocket-book for Chemists,' 'The Assay and Analysis of Iron and Steel, Iron Ores, and Fuel,' etc., etc. Royal 32mo, boards, gilt edges, 6s.

The Fireman's Guide; a Handbook on the Care of Boilers. By TEKNOLOG, föreningen T. I. Stockholm. Translated from the third edition, and revised by KARL P. DAHLSTROM, M.E. Second edition. Fcap. 8vo, cloth, 2s.

A Treatise on Modern Steam Engines and Boilers, including Land Locomotive, and Marine Engines and Boilers, for the use of Students. By FREDERICK COLYER, M. Inst. C.E., Mem. Inst. M.E. With 36 *plates.* 4to, cloth, 12s. 6d.

CONTENTS:

1. Introduction—2. Original Engines—3. Boilers—4. High-Pressure Beam Engines—5. Cornish Beam Engines—6. Horizontal Engines—7. Oscillating Engines—8. Vertical High-Pressure Engines—9. Special Engines—10. Portable Engines—11. Locomotive Engines—12. Marine Engines.

Steam Engine Management; a Treatise on the Working and Management of Steam Boilers. By F. COLYER, M. Inst. C.E., Mem. Inst. M.E. New edition, 18mo, cloth, 3s. 6d.

A Text-Book of Tanning, embracing the Preparation of all kinds of Leather. By HARRY R. PROCTOR, F.C.S., of Low Lights Tanneries. *With illustrations.* Crown 8vo, cloth, 10s. 6d.

Aid Book to Engineering Enterprise. By EWING MATHESON, M. Inst. C.E. The Inception of Public Works, Parliamentary Procedure for Railways, Concessions for Foreign Works, and means of Providing Money, the Points which determine Success or Failure, Contract and Purchase, Commerce in Coal, Iron, and Steel, &c. Second edition, revised and enlarged, 8vo, cloth, 21s.

Pumps, Historically, Theoretically, and Practically Considered. By P. R. BJÖRLING. With 156 *illustrations.* Crown 8vo, cloth, 7s. 6d.

The Marine Transport of Petroleum. A Book for the use of Shipowners, Shipbuilders, Underwriters, Merchants, Captains and Officers of Petroleum-carrying Vessels. By G. H. LITTLE, Editor of the 'Liverpool Journal of Commerce.' Crown 8vo, cloth, 10s. 6d.

Liquid Fuel for Mechanical and Industrial Purposes. Compiled by E. A. BRAYLEY HODGETTS. With *wood engravings.* 8vo, cloth, 7s. 6d.

Tropical Agriculture: A Treatise on the Culture, Preparation, Commerce and Consumption of the principal Products of the Vegetable Kingdom. By P. L. SIMMONDS, F.L.S., F.R.C.I. New edition, revised and enlarged, 8vo, cloth, 21s.

Health and Comfort in House Building; or, Ventilation with Warm Air by Self-acting Suction Power. With Review of the Mode of Calculating the Draught in Hot-air Flues, and with some Actual Experiments by J. DRYSDALE, M.D., and J. W. HAYWARD, M.D. *With plates and woodcuts.* Third edition, with some New Sections, and the whole carefully Revised, 8vo, cloth, 7s. 6d.

Losses in Gold Amalgamation. With Notes on the Concentration of Gold and Silver Ores. *With six plates.* By W. MCDERMOTT and P. W. DUFFIELD. 8vo, cloth, 5s.

A Guide for the Electric Testing of Telegraph Cables. By Col. V. HOSKIŒR, Royal Danish Engineers. Third edition, crown 8vo, cloth, 4s. 6d.

The Hydraulic Gold Miners' Manual. By T. S. G. KIRKPATRICK, M.A. Oxon. With 6 *plates.* Crown 8vo, cloth, 6s.

"We venture to think that this work will become a text-book on the important subject of which it treats. Until comparatively recently hydraulic mines were neglected. This was scarcely to be surprised at, seeing that their working in California was brought to an abrupt termination by the action of the farmers on the *débris* question, whilst their working in other parts of the world had not been attended with the anticipated success."—*The Mining World and Engineering Record.*

The Arithmetic of Electricity. By T. O'CONOR SLOANE. Crown 8vo, cloth, 4s. 6d.

The Turkish Bath: **Its Design and Construction for**
Public and Commercial Purposes. By R. O. ALLSOP, Architect. *With plans and sections.* 8vo, cloth, 6s.

Earthwork Slips and Subsidences upon Public Works:
Their Causes, Prevention and Reparation. Especially written to assist those engaged in the Construction or Maintenance of Railways, Docks, Canals, Waterworks, River Banks, Reclamation Embankments, Drainage Works, &c., &c. By JOHN NEWMAN, Assoc. Mem. Inst. C.E., Author of 'Notes on Concrete,' &c. Crown 8vo, cloth, 7s. 6d.

Gas and Petroleum Engines: **A Practical Treatise**
on the Internal Combustion Engine. By WM. ROBINSON, M.E., Senior Demonstrator and Lecturer on Applied Mechanics, Physics, &c., City and Guilds of London College, Finsbury, Assoc. Mem. Inst. C.E., &c. *Numerous illustrations.* 8vo, cloth, 14s.

Waterways and Water Transport in Different Countries. With a description of the Panama, Suez, Manchester, Nicaraguan, and other Canals. By J. STEPHEN JEANS, Author of 'England's Supremacy,' 'Railway Problems,' &c. *Numerous illustrations.* 8vo, cloth, 14s.

A Treatise on the Richards Steam-Engine Indicator
and the Development and Application of Force in the Steam-Engine. By CHARLES T. PORTER. Fourth Edition, revised and enlarged, 8vo, cloth, 9s.

CONTENTS.

The Nature and Use of the Indicator:
The several lines on the Diagram.
Examination of Diagram No. 1.
Of Truth in the Diagram.
Description of the Richards Indicator.
Practical Directions for Applying and Taking Care of the Indicator.
Introductory Remarks.
Units.
Expansion.
Directions for ascertaining from the Diagram the Power exerted by the Engine.
To Measure from the Diagram the Quantity of Steam Consumed.
To Measure from the Diagram the Quantity of Heat Expended.
Of the Real Diagram, and how to Construct it.
Of the Conversion of Heat into Work in the Steam-engine.
Observations on the several Lines of the Diagram.
Of the Loss attending the Employment of Slow-piston Speed, and the Extent to which this is Shown by the Indicator.
Of other Applications of the Indicator.
Of the use of the Tables of the Properties of Steam in Calculating the Duty of Boilers.
Introductory.
Of the Pressure on the Crank when the Connecting-rod is conceived to be of Infinite Length.
The Modification of the Acceleration and Retardation that is occasioned by the Angular Vibration of the Connecting-rod.
Method of representing the actual pressure on the crank at every point of its revolution.
The Rotative Effect of the Pressure exerted on the Crank.
The Transmitting Parts of an Engine, considered as an Equaliser of Motion.
A Ride on a Buffer-beam (Appendix).

In demy 4to, handsomely bound in cloth, *illustrated with* **220** *full page plates*,
Price 15s.

ARCHITECTURAL EXAMPLES
IN BRICK, STONE, WOOD, AND IRON.
A COMPLETE WORK ON THE DETAILS AND ARRANGEMENT OF BUILDING CONSTRUCTION AND DESIGN.

By WILLIAM FULLERTON, Architect.

Containing 220 Plates, with numerous Drawings selected from the Architecture of Former and Present Times.

The Details and Designs are Drawn to Scale, $\frac{1}{8}''$, $\frac{1}{4}''$, $\frac{1}{2}''$, and Full size being chiefly used.

The Plates are arranged in Two Parts. The First Part contains Details of Work in the four principal Building materials, the following being a few of the subjects in this Part:—Various forms of Doors and Windows, Wood and Iron Roofs, Half Timber Work, Porches, Towers, Spires, Belfries, Flying Buttresses, Groining, Carving, Church Fittings, Constructive and Ornamental Iron Work, Classic and Gothic Molds and Ornament, Foliation Natural and Conventional, Stained Glass, Coloured Decoration, a Section to Scale of the Great Pyramid, Grecian and Roman Work, Continental and English Gothic, Pile Foundations, Chimney Shafts according to the regulations of the London County Council, Board Schools. The Second Part consists of Drawings of Plans and Elevations of Buildings, arranged under the following heads:—Workmen's Cottages and Dwellings, Cottage Residences and Dwelling Houses, Shops, Factories, Warehouses, Schools, Churches and Chapels, Public Buildings, Hotels and Taverns, and Buildings of a general character.

All the Plates are accompanied with particulars of the Work, with Explanatory Notes and Dimensions of the various parts.

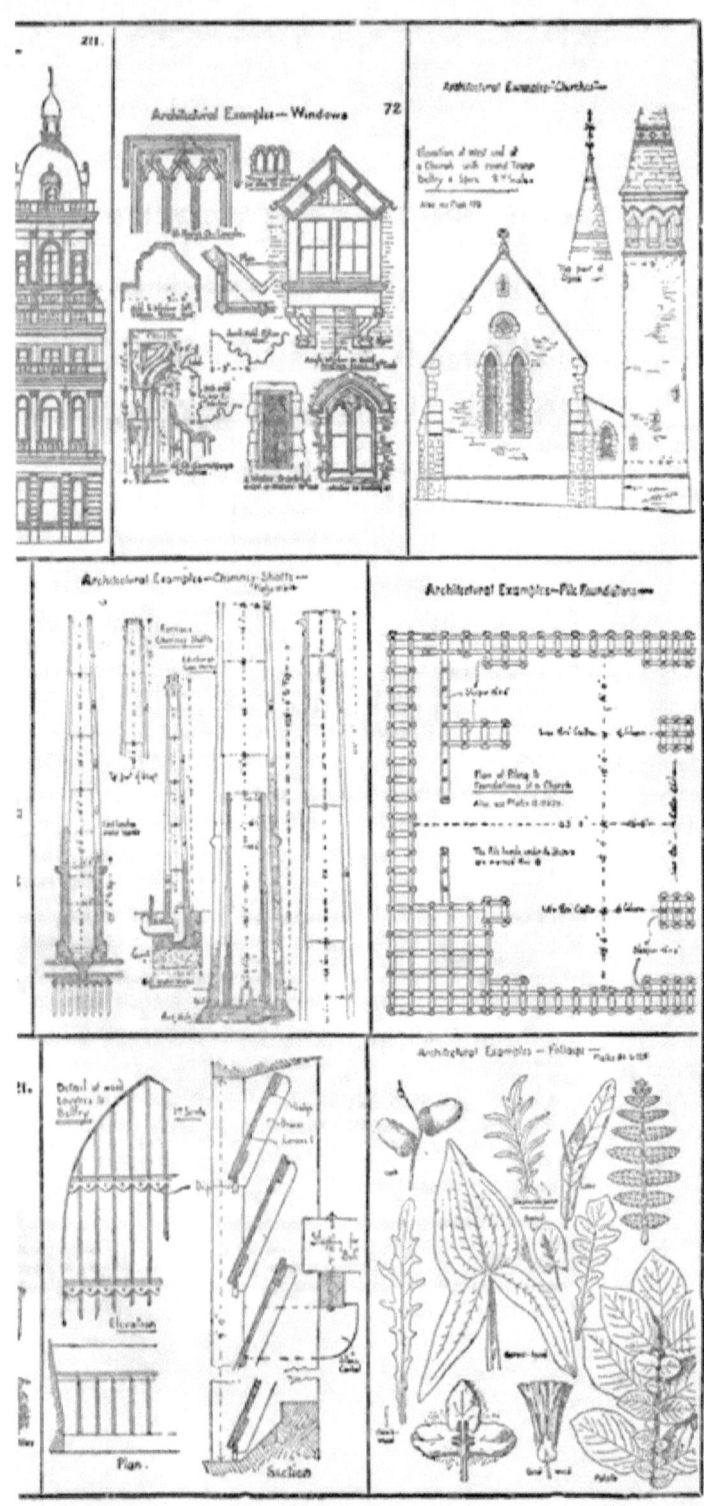

Crown 8vo, cloth, with illustrations, 5s.

WORKSHOP RECEIPTS,
FIRST SERIES.

By ERNEST SPON.

Synopsis of Contents.

Bookbinding.
Bronzes and Bronzing.
Candles.
Cement.
Cleaning.
Colourwashing.
Concretes.
Dipping Acids.
Drawing Office Details.
Drying Oils.
Dynamite.
Electro - Metallurgy — (Cleaning, Dipping, Scratch-brushing, Batteries, Baths, and Deposits of every description).
Enamels.
Engraving on Wood, Copper, Gold, Silver, Steel, and Stone.
Etching and Aqua Tint.
Firework Making — (Rockets, Stars, Rains, Gerbes, Jets, Tourbillons, Candles, Fires, Lances, Lights, Wheels, Fire-balloons, and minor Fireworks).
Fluxes.
Foundry Mixtures.
Freezing.
Fulminates.
Furniture Creams, Oils, Polishes, Lacquers, and Pastes.
Gilding.
Glass Cutting, Cleaning, Frosting, Drilling, Darkening, Bending, Staining, and Painting.
Glass Making.
Glues.
Gold.
Graining.
Gums.
Gun Cotton.
Gunpowder.
Horn Working.
Indiarubber.
Japans, Japanning, and kindred processes.
Lacquers.
Lathing.
Lubricants.
Marble Working.
Matches.
Mortars.
Nitro-Glycerine.
Oils.
Paper.
Paper **Hanging.**
Painting in **Oils,** in Water Colours, **as** well as Fresco, House, Transparency, Sign, and Carriage Painting.
Photography.
Plastering.
Polishes.
Pottery—(Clays, Bodies, Glazes, Colours, Oils, Stains, Fluxes, Enamels, and Lustres).
Scouring.
Silvering.
Soap.
Solders.
Tanning.
Taxidermy.
Tempering Metals.
Treating Horn, Mother-o'-Pearl, and like substances.
Varnishes, Manufacture and Use of.
Veneering.
Washing.
Waterproofing.
Welding.

Besides Receipts relating to the lesser Technological matters **and** processes, such as the manufacture and use of Stencil Plates, Blacking, Crayons, Paste, Putty, Wax, Size, Alloys, Catgut, Tunbridge Ware, Picture Frame and Architectural Mouldings, Compos, Cameos, and others **too** numerous to mention.

PUBLISHED BY E. & F. N. SPON. 23

Crown 8vo, cloth, 485 pages, with illustrations, 5s.

WORKSHOP RECEIPTS,
SECOND SERIES.
By ROBERT HALDANE.

SYNOPSIS OF CONTENTS.

Acidimetry and Alkalimetry.	Disinfectants.	Iodoform.
Albumen.	Dyeing, Staining, and Colouring.	Isinglass.
Alcohol.	Essences.	Ivory substitutes.
Alkaloids.	Extracts.	Leather.
Baking-powders.	Fireproofing.	Luminous bodies.
Bitters.	Gelatine, Glue, and Size.	Magnesia.
Bleaching.	Glycerine.	Matches.
Boiler Incrustations.	Gut.	Paper.
Cements and Lutes.	Hydrogen peroxide.	Parchment.
Cleansing.	Ink.	Perchloric acid.
Confectionery.	Iodine.	Potassium oxalate.
Copying.		Preserving.

Pigments, Paint, and Painting: embracing the preparation of *Pigments*, including alumina lakes, blacks (animal, bone, Frankfort, ivory, lamp, sight, soot), blues (antimony, Antwerp, cobalt, cœruleum, Egyptian, manganate, Paris, Péligot, Prussian, smalt, ultramarine), browns (bistre, hinau, sepia, sienna, umber, Vandyke), greens (baryta, Brighton, Brunswick, chrome, cobalt, Douglas, emerald, manganese, mitis, mountain, Prussian, sap, Scheele's, Schweinfurth, titanium, verdigris, zinc), reds (Brazilwood lake, carminated lake, carmine, Cassius purple, cobalt pink, cochineal lake, colcothar, Indian red, madder lake, red chalk, red lead, vermilion), whites (alum, baryta, Chinese, lead sulphate, white lead—by American, Dutch, French, German, Kremnitz, and Pattinson processes, precautions in making, and composition of commercial samples—whiting, Wilkinson's white, zinc white), yellows (chrome, gamboge, Naples, orpiment, realgar, yellow lakes); *Paint* (vehicles, testing oils, driers, grinding, storing, applying, priming, drying, filling, coats, brushes, surface, water-colours, removing smell, discoloration; miscellaneous paints—cement paint for carton-pierre, copper paint, gold paint, iron paint, lime paints, silicated paints, steatite paint, transparent paints, tungsten paints, window paint, zinc paints); *Painting* (general instructions, proportions of ingredients, measuring paint work; carriage painting—priming paint, best putty, finishing colour, cause of cracking, mixing the paints, oils, driers, and colours, varnishing, importance of washing vehicles, re-varnishing, how to dry paint; woodwork painting).

Crown 8vo, cloth, 480 pages, with 183 illustrations, 5s.

WORKSHOP RECEIPTS,

THIRD SERIES.

By C. G. WARNFORD LOCK.

Uniform with the First and Second Series.

SYNOPSIS OF CONTENTS.

Alloys.	Indium.	Rubidium.
Aluminium.	Iridium.	Ruthenium.
Antimony.	Iron and Steel.	Selenium.
Barium.	Lacquers and Lacquering.	Silver.
Beryllium.	Lanthanum.	Slag.
Bismuth.	Lead.	Sodium.
Cadmium.	Lithium.	Strontium.
Cæsium.	Lubricants.	Tantalum.
Calcium.	Magnesium.	Terbium.
Cerium.	Manganese.	Thallium.
Chromium.	Mercury.	Thorium.
Cobalt.	Mica.	Tin.
Copper.	Molybdenum.	Titanium.
Didymium.	Nickel.	Tungsten.
Electrics.	Niobium.	Uranium.
Enamels and Glazes.	Osmium.	Vanadium.
Erbium.	Palladium.	Yttrium.
Gallium.	Platinum.	Zinc.
Glass.	Potassium.	Zirconium.
Gold.	Rhodium.	

WORKSHOP RECEIPTS,
FOURTH SERIES,
DEVOTED MAINLY TO HANDICRAFTS & MECHANICAL SUBJECTS.

By C. G. WARNFORD LOCK.

250 Illustrations, with Complete Index, and a General Index to the Four Series, 5s.

Waterproofing — rubber goods, cuprammonium processes, miscellaneous preparations.

Packing and Storing articles of delicate odour or colour, of a deliquescent character, liable to ignition, apt to suffer from insects or damp, or easily broken.

Embalming and Preserving anatomical specimens.

Leather Polishes.

Cooling Air and Water, producing low temperatures, making ice, cooling syrups and solutions, and separating salts from liquors by refrigeration.

Pumps and Siphons, embracing every useful contrivance for raising and supplying water on a moderate scale, and moving corrosive, tenacious, and other liquids.

Desiccating—air- and water-ovens, and other appliances for drying natural and artificial products.

Distilling—water, tinctures, extracts, pharmaceutical preparations, essences, perfumes, and alcoholic liquids.

Emulsifying as required by pharmacists and photographers.

Evaporating—saline and other solutions, and liquids demanding special precautions.

Filtering—water, and solutions of various kinds.

Percolating and Macerating.

Electrotyping.

Stereotyping by both plaster and paper processes.

Bookbinding in all its details.

Straw Plaiting and the fabrication of baskets, matting, etc.

Musical Instruments—the preservation, tuning, and repair of pianos, harmoniums, musical boxes, etc.

Clock and Watch Mending—adapted for intelligent amateurs.

Photography—recent development in rapid processes, handy apparatus, numerous recipes for sensitizing and developing solutions, and applications to modern illustrative purposes.

NOW COMPLETE.

With nearly 1500 *illustrations*, in super-royal 8vo, in 5 Divisions, cloth. Divisions 1 to 4, 13s. 6d. each; Division 5, 17s. 6d.; or 2 vols., cloth, £3 10s.

SPONS' ENCYCLOPÆDIA
OF THE
INDUSTRIAL ARTS, MANUFACTURES, AND COMMERCIAL PRODUCTS.

EDITED BY C. G. WARNFORD LOCK, F.L.S.

Among the more important of the subjects treated of, are the following:—

Acids, 207 pp. 220 figs.
Alcohol, 23 pp. 16 figs.
Alcoholic Liquors, 13 pp.
Alkalies, 89 pp. 78 figs.
Alloys. Alum.
Asphalt. Assaying.
Beverages, 89 pp. 29 figs.
Blacks.
Bleaching Powder, 15 pp.
Bleaching, 51 pp. 48 figs.
Candles, 18 pp. 9 figs.
Carbon Bisulphide.
Celluloid, 9 pp.
Cements. Clay.
Coal-tar Products, 44 pp. 14 figs.
Cocoa, 8 pp.
Coffee, 32 pp. 13 figs.
Cork, 8 pp. 17 figs.
Cotton Manufactures, 62 pp. 57 figs.
Drugs, 38 pp.
Dyeing and Calico Printing, 28 pp. 9 figs.
Dyestuffs, 16 pp.
Electro-Metallurgy, 13 pp.
Explosives, 22 pp. 33 figs.
Feathers.
Fibrous Substances, 92 pp. 79 figs.
Floor-cloth, 16 pp. 21 figs.
Food Preservation, 8 pp.
Fruit, 8 pp.

Fur, 5 pp.
Gas, Coal, 8 pp.
Gems.
Glass, 45 pp. 77 figs.
Graphite, 7 pp.
Hair, 7 pp.
Hair Manufactures.
Hats, 26 pp. 26 figs.
Honey. Hops.
Horn.
Ice, 10 pp. 14 figs.
Indiarubber Manufactures, 23 pp. 17 figs.
Ink, 17 pp.
Ivory.
Jute Manufactures, 11 pp., 11 figs.
Knitted Fabrics — Hosiery, 15 pp. 13 figs.
Lace, 13 pp. 9 figs.
Leather, 28 pp. 31 figs.
Linen Manufactures, 16 pp. 6 figs.
Manures, 21 pp. 30 figs.
Matches, 17 pp. 38 figs.
Mordants, 13 pp.
Narcotics, 47 pp.
Nuts, 10 pp.
Oils and Fatty Substances, 125 pp.
Paint.
Paper, 26 pp. 23 figs.
Paraffin, 8 pp. 6 figs.
Pearl and Coral, 8 pp.
Perfumes, 10 pp.

Photography, 13 pp. 20 figs.
Pigments, 9 pp. 6 figs.
Pottery, 46 pp. 57 figs.
Printing and Engraving, 20 pp. 8 figs.
Rags.
Resinous and Gummy Substances, 75 pp. 16 figs.
Rope, 16 pp. 17 figs.
Salt, 31 pp. 23 figs.
Silk, 8 pp.
Silk Manufactures, 9 pp. 11 figs.
Skins, 5 pp.
Small Wares, 4 pp.
Soap and Glycerine, 39 pp. 45 figs.
Spices, 16 pp.
Sponge, 5 pp.
Starch, 9 pp. 10 figs.
Sugar, 155 pp. 134 figs.
Sulphur.
Tannin, 18 pp.
Tea, 12 pp.
Timber, 13 pp.
Varnish, 15 pp.
Vinegar, 5 pp.
Wax, 5 pp.
Wool, 2 pp.
Woollen Manufactures, 58 pp. 39 figs.

In super-royal 8vo, 1168 pp., *with* 2400 *illustrations*, in 3 Divisions, cloth, price 13s. 6d. each; or 1 vol., cloth, 2l.; or half-morocco, 2l. 8s.

A SUPPLEMENT

TO

SPONS' DICTIONARY OF ENGINEERING.

Edited by ERNEST SPON, Memb. Soc. Engineers.

Abacus, Counters, Speed Indicators, and Slide Rule.
Agricultural Implements and Machinery.
Air Compressors.
Animal Charcoal Machinery.
Antimony.
Axles and Axle-boxes.
Barn Machinery.
Belts and Belting.
Blasting. Boilers.
Brakes.
Brick Machinery.
Bridges.
Cages for Mines.
Calculus, Differential and Integral.
Canals.
Carpentry.
Cast Iron.
Cement, Concrete, Limes, and Mortar.
Chimney Shafts.
Coal Cleansing and Washing.
Coal Mining.
Coal Cutting Machines.
Coke Ovens. Copper.
Docks. Drainage.
Dredging Machinery.
Dynamo - Electric and Magneto-Electric Machines.
Dynamometers.
Electrical Engineering, Telegraphy, Electric Lighting and its practical details, Telephones
Engines, Varieties of.
Explosives. Fans.
Founding, Moulding and the practical work of the Foundry.
Gas, Manufacture of.
Hammers, Steam and other Power.
Heat. Horse Power.
Hydraulics.
Hydro-geology.
Indicators. Iron.
Lifts, Hoists, and Elevators.
Lighthouses, Buoys, and Beacons.
Machine Tools.
Materials of Construction.
Meters.
Ores, Machinery and Processes employed to Dress.
Piers.
Pile Driving.
Pneumatic Transmission.
Pumps.
Pyrometers.
Road Locomotives.
Rock Drills.
Rolling Stock.
Sanitary Engineering.
Shafting.
Steel.
Steam **Navvy**.
Stone Machinery.
Tramways.
Well Sinking.

JUST PUBLISHED.

In demy 8vo, cloth, 600 pages, and 1420 Illustrations, 6s.

SPONS'
MECHANICS' OWN BOOK;
A MANUAL FOR HANDICRAFTSMEN AND AMATEURS.

CONTENTS.

Mechanical Drawing—Casting and Founding in Iron, Brass, Bronze, and other Alloys—Forging and Finishing Iron—Sheetmetal Working—Soldering, Brazing, and Burning—Carpentry and Joinery, embracing descriptions of some 400 Woods, over 200 Illustrations of Tools and their uses, Explanations (with Diagrams) of 116 joints and hinges, and Details of Construction of Workshop appliances, rough furniture, Garden and Yard Erections, and House Building—Cabinet-Making and Veneering—Carving and Fretcutting—Upholstery—Painting, Graining, and Marbling—Staining Furniture, Woods, Floors, and Fittings—Gilding, dead and bright, on various grounds—Polishing Marble, Metals, and Wood—Varnishing—Mechanical movements, illustrating contrivances for transmitting motion—Turning in Wood and Metals—Masonry, embracing Stonework, Brickwork, Terracotta, and Concrete—Roofing with Thatch, Tiles, Slates, Felt, Zinc, &c.—Glazing with and without putty, and lead glazing—Plastering and Whitewashing—Paper-hanging—Gas-fitting—Bell-hanging, ordinary and electric Systems—Lighting—Warming—Ventilating—Roads, Pavements, and Bridges—Hedges, Ditches, and Drains—Water Supply and Sanitation—Hints on House Construction suited to new countries.

E. & F. N. SPON, 125, Strand, London.
New York: 12, Cortlandt Street.

www.ingramcontent.com/pod-product-compliance
Lightning Source LLC
Chambersburg PA
CBHW020126170426
43199CB00009B/663